W9-ADW-330

Anatomy and Physiology

LABORATORY MANUAL

NINTH EDITION

Anatomy and Physiology

LABORATORY MANUAL

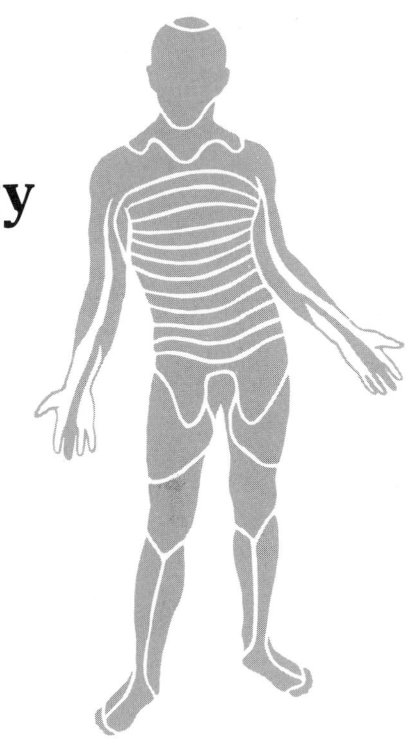

CATHERINE PARKER ANTHONY, R.N., B.A., M.S.

Formerly Assistant Professor of Nursing, Science Department, and Assistant Instructor of Anatomy and Physiology, Frances Payne Bolton School of Nursing, Case Western Reserve University, Cleveland, Ohio; formerly Instructor of Anatomy and Physiology, Lutheran Hospital and St. Luke's Hospital, Cleveland, Ohio

With 115 drawings, 69 to be labeled

The C. V. Mosby Company

Saint Louis 1975

Ninth edition

Previous editions copyrighted 1944, 1946, 1949, 1955, 1959, 1963, 1967, 1971
Printed in the United States of America
International Standard Book Number 0-8016-0269-6
Distributed in Great Britain by Henry Kimpton, London

TS/CB/B 9 8 7 6 5 4 3 2 1

PREFACE

This ninth edition of *Anatomy and Physiology Laboratory Manual* is offered as a guide to help students use laboratory hours for exploring, discovering, clarifying, and verifying information about the human body. Most of the procedures require students to work independently with only minimal assistance from the instructor. All of the procedures pose problems and clearly indicate the steps to be taken for solving them by the scientific method. To further assure understanding and appreciation of this method—that discovers "textbook facts" and that so largely has shaped our civilization—many fine films are suggested. Films make it possible for students to see how research scientists use the scientific method to investigate some of the still uncertain workings of the body.

The major aim of the procedures in this manual is to help students learn significant information about the body rather than laboratory techniques and how to use complex equipment. Hence both students and instructors are spared much tedious, time-consuming work before, during, and after laboratory sessions.

Some procedures have been revised in this edition, a few have been deleted, and several new ones have been added. Examples of the latter are the procedures on respiratory volumes and capacities, estimation of venous pressure, and blood typing. Almost every chapter contains new film titles. A new chapter on the reproduction of cells has been added. The format that permits rapid checking of students' conclusions and the self-tests (somewhat modified) have been retained.

CATHERINE PARKER ANTHONY

Outline for 90-hour course

Following is a suggested division of hours for a 90-hour course. The instructor may, of course, vary these hours in many ways. The outline is offered merely as a suggestion of possible time allotment.

		Suggested time allotment			
		With maximum laboratory hours		*With minimum laboratory hours*	
Subject	*Total hours**	*Lecture (hr)*	*Laboratory (hr)*	*Lecture (hr)*	*Laboratory (hr)*
Unit one					
Organization of the body	2	None	2	None	2
Cells	4	2	2	2	2
Unit two					
The skeletal system	7	2	5	5	2
The muscular system	8	2	6	6	2
Unit three					
The somatic nervous system, the autonomic nervous system, and sense organs	18	6	12	12	6
The endocrine system	3	2	1	2	1
Unit four					
The respiratory system	5	2	3	3	2
The cardiovascular system	17	5	12	12	5
The digestive system and metabolism	13	4	9	9	4
The urinary system	5	2	3	3	2
Unit five					
Reproduction of cells and reproduction	8	3	5	5	3
Total	90	30	60	61	29

*For a 96-hour course, 2 more hours might be allowed for Unit one, 2 more for the muscular system (Unit two), and 2 more for reproduction (Unit five). For a 100-hour course, 2 more hours might be allowed for Unit one, 2 more for the muscular system (Unit two), 2 more for the somatic nervous system, the autonomic nervous system, and sense organs (Unit three), 1 more for the cardiovascular system (Unit four), 1 more for the respiratory system (Unit four), and 2 more for reproduction (Unit five).

CONTENTS

UNIT FOUR

MAINTAINING THE METABOLISM OF THE BODY

UNIT FIVE

REPRODUCTION

UNIT SIX

Suggestions to students

Laboratory work is included in anatomy and physiology courses to help you in your efforts to increase your understanding of the human body. Whether you learn much or little from laboratory sessions depends mainly upon the type of mental habits you cultivate. The following suggestions are offered in the hope that they may help you to gain the maximum benefit from your laboratory work.

1 Come to the laboratory with an attitude of expectancy and with a conscious realization that the procedures are designed to serve two purposes—to demonstrate facts and principles discussed in the textbook or lecture and to provide experiences that will help you to learn the most important facts of the course. (In this sense, the laboratory period is a planned study period.)

2 When examining fresh or preserved anatomical specimens, think of them as materials from which you can learn, at first hand, facts about which you have read or heard. Do not think of them as parts of a person's or an animal's body. In other words, cultivate an attitude of intellectual curiosity and suppress attitudes of morbid curiosity. If you conscientiously cultivate this scientific method of approach to dissection and demonstrations, you will never be distressed by any specimens that you may be called upon to handle or witness. It is of paramount importance for a student to develop this professional attitude of intellectual curiosity about things that she or he sees.

3 Be quiet in manner and orderly in your habits in the laboratory. Always leave your laboratory unit clean and neat.

4 Handle equipment very carefully, since it is costly and often difficult to replace.

Laboratory equipment

General

1. Dissectible torso model of human body
2. Model of eye
3. Model of ear
4. Model of section of spinal cord
5. Articulated skeleton
6. Dissectible skull
7. Disarticulated bones
8. Prepared tissue and membrane slides
9. Artificial cells (parchment or collodion)
10. Hemocytometer, counting slide, and mixing pipettes
11. Tallqvist scale for hemoglobin estimation
12. Set of standard anatomical charts
13. Covered glass specimen jar
14. Absorbent cotton
15. Beakers and test tubes
16. Ring stands
17. Stethoscope
18. Sphygmomanometer

Preserved specimens (if available)

1. Human brain
2. Human spinal cord
3. Human stomach
4. Human gallbladder
5. Human heart
6. Human kidney
7. Human uterus, tubes, and ovary
8. Human embryos
9. Embalmed cat

Fresh materials

1. Beef joint sawed longitudinally
2. Two long beef bones—one sawed longitudinally and one sawed horizontally
3. Yellow or white onion
4. Sheep brains
5. Sheep pluck
6. Sheep hearts
7. Sheep spinal cords
8. Sheep kidneys
9. Two chicken legs
10. Beef eyes
11. Several live rats, frogs, or guinea pigs
12. Sheep liver and gallbladder

Solutions

1. 10% glucose
2. 5% or 2% glucose
3. 10% formalin
4. Methylene blue
5. Acetic acid
6. 10% urethane
7. Ether
8. Thin, cooked starch
9. Benedict's solution
10. Tincture of iodine
11. Tincture of cantharides

Equipment for each two students

1. Dissecting kit containing the following:
   - **a** Scalpel
   - **b** Scissors
   - **c** Thumb forceps
   - **d** Medicine dropper
   - **e** Blunt probe
   - **f** Dissecting needle
2. Shallow dissecting pan
3. Microscope

Equipment to be provided by each student

1. Box of colored pencils
2. Hard lead (4H) drawing pencil for labeling and shading (does not smear)
3. Eraser
4. Ruler

Reference books for use in laboratory

1. Atlases on anatomy
2. Textbooks of anatomy and physiology
3. Textbooks of histology
4. Textbooks of neuroanatomy
5. Textbooks of embryology

Sources from which laboratory supplies may be secured

1. Models of torso, eye, ear, etc. (also skeleton)—any biological supply house, such as Clay-Adams Co., Inc., 141 E. 25th St., New York, N.Y. 10010
2. Preserved human specimens—your hospital department of pathology
3. Fresh materials, such as sheep brains and beef joints—your local slaughterhouse or butcher
4. Live rats and frogs and embalmed animals—any biological supply house
5. Solutions—your hospital pharmacy
6. Dissecting kits, beakers, test tubes, and bell jar—any chemical supply house
7. Recording systems and teaching kits—Harvard Apparatus Co., Inc., Smith Street, Dover, Mass. 02030
8. Blood typing—Lab-Aids, Inc., 160 Rome St., Farmingdale, N.Y. 11735

UNIT ONE

THE BODY AS A WHOLE

1 Organization of the body

Structural organization of body

Procedure A—Examination of human torso model

Equipment

1 Human torso model
2 Anatomical charts and illustrations
3 Textbooks of anatomy and physiology

Problems

1 Where are the four main body cavities located?
2 What organs are located in the thoracic cavity (exclusive of blood vessels, lymphatic vessels, and nerves)?
3 What organs are located in each of the nine regions of the abdominal portion of the abdominopelvic cavity?

Collection of data

1 Consult a textbook to find the names and locations of the four main body cavities.
2 Examine the color insert on human anatomy and Fig. 1-3 in Anthony, C. P., and Kolthoff, N. J.: Textbook of anatomy and physiology, ed. 9, St. Louis, 1975, The C. V. Mosby Co.
3 Remove the organs from the torso model. As you do so, check the name of each organ (in the color insert in the textbook or in other illustrations). Note the location, size, and shape of each organ as you remove it from the model. After you have removed all organs from the model, return them to their proper locations.

Conclusions

1 Label Fig. 1-1 as directed.

2 Label Fig. 1-2 as directed.

3 Of the following organs, which ones are located in the right hypochondriacal region of the abdominal cavity?

appendix	liver
ascending colon	pancreas
descending colon	spleen
gallbladder	stomach

____

4 Which, if any, of the above organs are located in the epigastrium?

____

5 Which, if any, are located in the right iliac region?

____

6 Which, if any, are located in the hypogastric region?

____

7 In which region is the spleen located?

____

8 Which, if any, of the above organs is located in the left iliac region?

____

9 Notable characteristics of the human body's architectural plan are a backbone, bisymmetry, and, as you have observed, cavities containing numerous organs.

a The diaphragm muscle forms the floor of a cavity located inside the rib cage. The name of this cavity is the ....

____

b The large cavity below the diaphragm is named the ....

____

c What organ occupies almost all of the space in the cranial cavity?

____

d In what cavity is the spinal cord located?

____

e Where are the internal reproductive organs located?

____

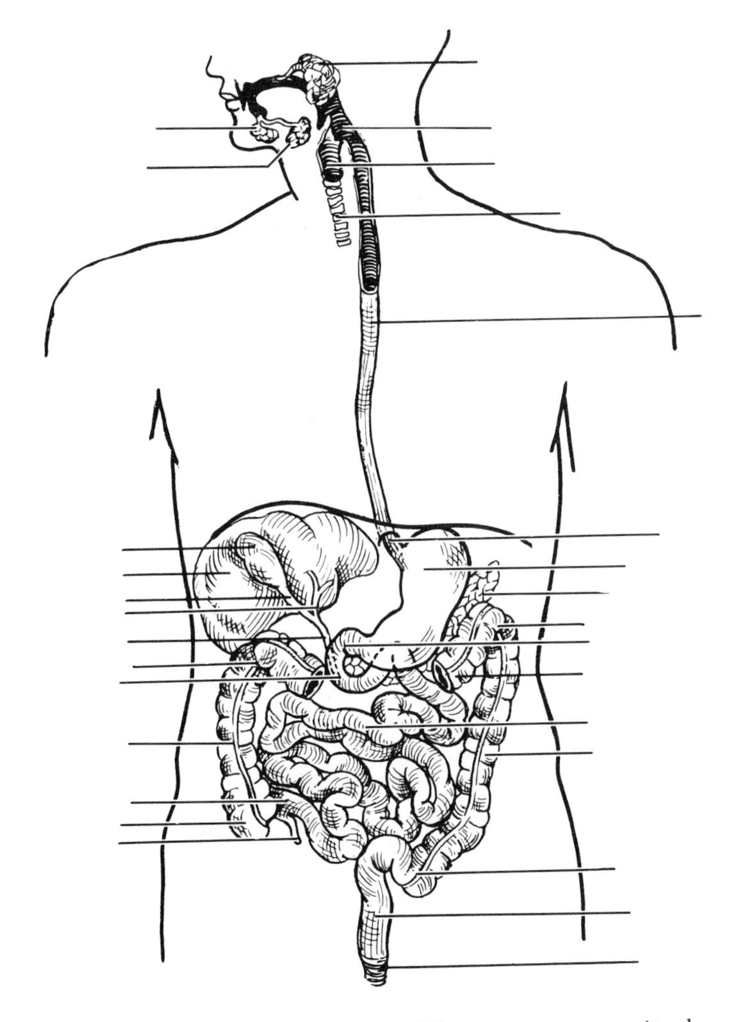

Fig. 1-1 Gastrointestinal tract. Place each of the following terms opposite the appropriate label line after first identifying each structure on the torso model of the body:

Anus
Ascending colon
Cecum
Common bile duct
Cystic duct
Descending colon
Duodenum
Esophagus
Gallbladder
Hepatic duct
Hepatic flexure
Ileum
Jejunum
Larynx
Liver
Pancreas
Parotid gland and duct
Pharynx
Rectum
Region of cardiac sphincter
Region of ileocecal sphincter
Region of pyloric sphincter
Sigmoid colon
Splenic flexure
Stomach
Sublingual gland
Submaxillary gland
Trachea
Transverse colon
Vermiform appendix

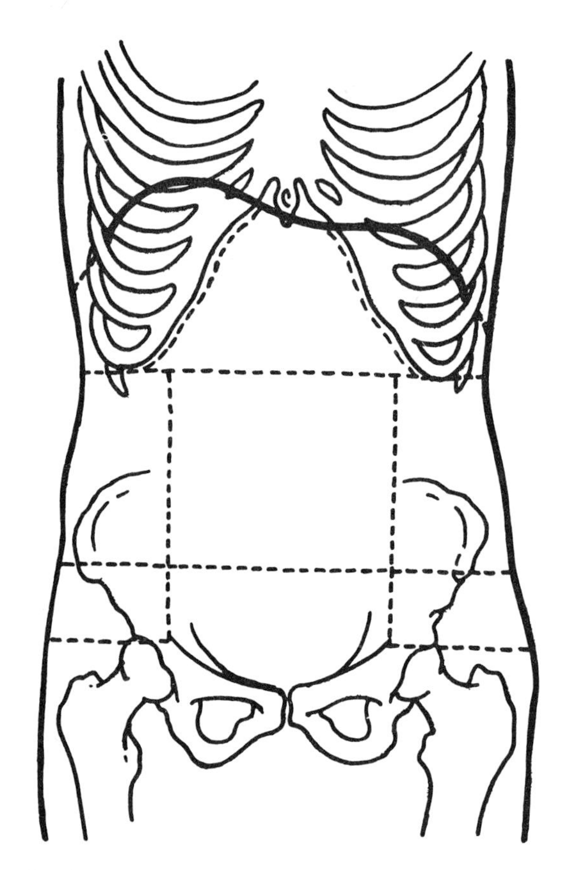

Fig. 1-2 Abdominal regions. Label the following:

Epigastric region
Hypogastric region
Left hypochondriac region
Left iliac region
Left lumbar region
Umbilical region

Structural organization of body—cont'd

Procedure B—Dissection of small animal

Equipment

1 Small mammal (e.g., rat, guinea pig, or cat)
2 Large bell jar
3 Ether
4 Cotton
5 Dissecting instruments
6 Dissecting board

Problems

1 How does the internal structural organization of this animal compare with that of the human being?
2 What are the main gross characteristics of living tissues and organs?

Collection of data

1 Anesthetize animal.
  a Cover the animal with a large bell jar. Slip a large piece of cotton saturated with ether under the jar. The animal will become excited for a brief period but then will quiet down.
  b Test to see whether the animal has entered the stage of surgical anesthesia by giving it a sudden prod with the side of the bell jar. If it does not respond, you may start dissection.
2 Dissect animal.
  a Place the animal ventral side up on a dissecting board.
  b Put the ether-soaked piece of cotton over the animal's nostrils and mouth to keep it anesthetized while you are performing your dissection—but watch the color of its lips and nostrils. If you see their normal red color fading to pink or white, take the ether-soaked cotton away. Replace it, freshly soaked with ether, the moment the animal shows any signs of coming out of anesthesia.
  c Make a midline incision through the skin from the pubis to the sternum.
  d Cut through the midline of the internal layer of the abdominal wall, exercising care not to cut into the underlying organs.
  e Make a second incision, starting again from the pubis but this time cutting obliquely forward and to the right.
  f Make a similar cut to the left. Make these incisions long enough to fold back the abdominal wall to expose the viscera.
3 Observe the following:
  a Greater omentum, the fatty apronlike structure lying over the intestines and attached at its upper border to the transverse colon and some other structures
  b Arrangement of the organs in the abdominal cavity; identify each organ
  c Smooth, shiny, serous lining of the abdominal cavity
  d Smooth, shiny, serous covering adherent to the abdominal organs
  e Color, texture, moistness, and pliability of the organs
4 Remove the organs from the abdominal cavity.
5 Observe the organs in the pelvic cavity.
6 Open the thoracic cavity by extending the midline incision through the breastbone and making a cut to the right and one to the left from the midline just above the diaphragm.
7 Observe:
  a Smooth, shiny, serous membrane lining of the thoracic cavity
  b Sac enclosing the heart
  c Domelike shape of the diaphragm muscle separating the thoracic cavity from the abdominal cavity
  d Thoracic organs; identify each organ
8 Sever the great aorta in order to kill the animal by bleeding.

Conclusions

1 From your own observations of the human torso model and of the internal structural arrangement of this small mammal, do you conclude that the latter is very different or very similar to that of the human body?

____

2 Based on your observation of the tissues of this freshly killed animal, do you think living tissue would be accurately described as:

a Dull or shiny?

____

b Dry or moist?

____

c Smooth or rough?

____

Structural organization of body—cont'd

Procedure C—Films

"Incredible voyage," color, sound, 16 mm, 26 min; McGraw-Hill Films, Dept. WP, 330 West 42nd St., New York, N.Y. 10036 (purchase, $385; rental, $18). By means of special instruments and photographic techniques, viewer is taken on a conducted tour through the living body.

and/or

"Corps profound," color, 16 mm, sound, 22 min; Association Films, 600 Grand Ave., Ridgefield, N.J. 07657 (free loan). Shows many body parts, including pituitary gland, middle ear, and abdominal organs.

Self-test

1 If you looked in the right hypochondriacal region of the abdominal cavity, what organ would you see?

____

2 If you looked in the left hypochondriacal region of the abdominal cavity, what organ would you see?

____

3 If you looked in the umbilical region of the abdominal cavity, what organ would you see?

____

4 In what body cavity would you look to see the ovaries?

____

5 Locate the pancreas with reference to the stomach.

____

6 Locate the urinary bladder with reference to the uterus.

____

7 Removal of the spleen would? would not? be an example of thoracic surgery.

____

8 The esophagus extends from the pharynx to what organ?

____

9 The trachea extends from the larynx to what organs?

____

10 In what portion of what cavity is the thymus located?

____

2 Cells

■ Cell's chemical structure

Procedure—Film

"A cell's chemical organization," color, 16 mm, 16 min; McGraw-Hill Films, Dept. WP, 330 West 42nd St., New York, N.Y. 10036 (purchase, $215; rental, $12.50).

■ Movement of substances through cell membranes

Procedure A—Experiments demonstrating diffusion

Equipment

1. Albumin (uncooked white of egg)
2. Beakers
3. Dialyzing membrane (cellophane bag)
4. Forceps
5. Test tubes
6. Benedict's solution
7. Collodion
8. Distilled water
9. Gelatin
10. 5% glucose solution
11. Lugol's iodine solution
12. Silver nitrate solution
13. Sodium chloride
14. Potassium permanganate tablets

Problems

1. What do the terms diffusion, net diffusion, and dialysis mean?
2. Do solutes diffuse through water?
3. Can diffusion occur through a colloidal solution when it is in the semisolid gel state?
4. Do the crystalloids glucose and sodium chloride diffuse through a cellophane membrane?
5. Does the colloid albumin (egg white) diffuse through a cellophane membrane?
6. In which direction does net diffusion of a substance occur?

Collection of data

1. Preliminary preparation (a day or two before this laboratory period): add 5 gm (or 1 tsp) of gelatin to 25 ml of cold water. Let stand 5 minutes. Add 75 ml of boiling water and stir until dissolved. Pour about 10 ml of the solution into a test tube. Fill another test tube with the solution. Refrigerate until gelled.
2. Place a drop of methylene blue or some other dye on the surface of the gel prepared ahead of time. Set aside at room temperature. At the end of 1 hour, 2 hours, and 24 hours, observe whether the dye has moved and, if so, in which direction(s) and how far.
3. Place a few crystals of potassium permanganate on the bottom of a beaker half filled with water.

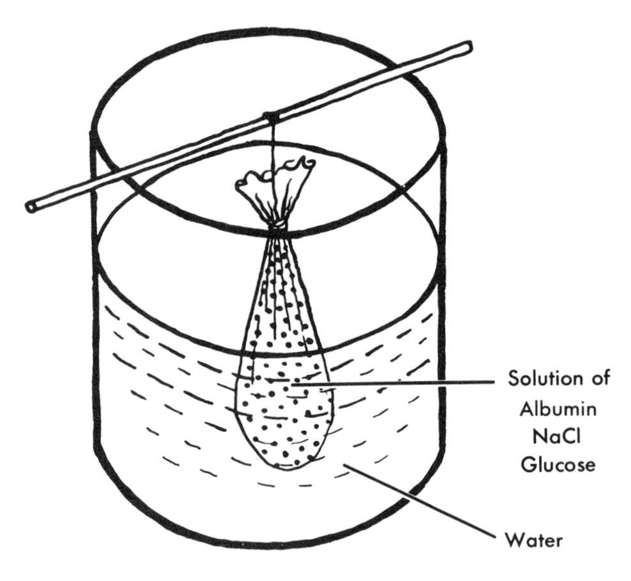

Fig. 2-1 Diffusion setup.

4 Prepare a bag from a 9 × 9 inch piece of cellophane (see Fig. 2-1).

5 Prepare a solution as follows: to 100 ml of water, add ½ tsp of sodium chloride, 50 ml of 5% glucose, and the uncooked white of an egg. Both sodium chloride and glucose are crystalloids or true solutes, whereas egg white or albumin is a colloidal solute. (Better look these terms up if you are not sure what they mean.)

6 Pour the solution you prepared in step 5 into the cellophane bag you made in step 4. Tie a string around the top of the bag and suspend it in distilled water as shown in Fig. 2-1. Let stand for about 1 hour, and then perform the tests indicated in steps 7, 8, and 9.

7 *Test for albumin:* Pour about 5 ml of the fluid in the beaker into a test tube. Add a few drops of nitric acid. Note whether coagulation occurs. Nitric acid coagulates albumin.

8 *Test the water in the beaker for sodium chloride:* Pour about 5 ml of the fluid in the beaker into a test tube and add a drop of silver nitrate. Note whether a precipitate forms. If sodium chloride is present, it will combine with silver nitrate to form a precipitate of silver chloride.

9 *Test the water in the beaker for glucose:* Put 5 ml of Benedict's solution into a test tube and add 4 or 5 drops of the beaker water. Boil 2 minutes. Cool slowly. Note whether a green, yellow, or red precipitate forms, indicating the presence of glucose.

10 Look up the words diffusion and dialysis in a dictionary and in a chemistry or physiology textbook.

Conclusions

1 Did the dye diffuse through the colloid, gelatin, in its gel state?

____

If your answer is no, omit the next two items, but if it is yes, answer them.

2 In which directions did the dye move through the gel?

____

3 How many centimeters (cm) do you estimate the dye moved from its original location on the gel in 24 hours? (1 cm is slightly less than ½ inch.) 1 cm or less? More than 1 cm but less than 5 cm? More than 5 cm?

____

4 Cytoplasm contains colloids (proteins, mainly) and may exist in the gel state. Based on the results you observed in step 2 under collection of data, do you postulate that substances can move through cytoplasm by the process of diffusion?

____

5 Did the results you observed in step 3 under collection of data indicate that potassium permanganate diffused through water?

____

6 Based on the results you observed in steps 2 and 3 under collection of data, do you postulate that diffusion occurs more rapidly through a gel, or a liquid, or at about the same rate through both?

___

7 Did the results observed in steps 7, 8, and 9 under collection of data indicate that the crystalloids glucose and sodium chloride diffused through the membrane used in step 6?

___

8 Did the colloidal solute albumin diffuse through the membrane used in this experiment (step 6 under collection of data)?

___

9 Diffusion is one of the physical processes by which water and? but not? solutes can pass through membranes permeable to them.

___

10 What term means the separation of crystalloids from colloids by diffusion of crystalloids through a membrane permeable to them but impermeable to colloids?

___

11 Which experiments under collection of data, if any, demonstrated dialysis: step 2? step 3? steps 5 and 6?

___

12 What does net diffusion mean?

___

13 The experiments you performed in steps 2 and 3 demonstrated that net diffusion of the solutes occurred from the site of the solute's greater? lesser? concentration to that of its greater? lesser? concentration.

___

___

14 The experiments you performed in steps 2 and 3 demonstrated the first law of diffusion—that net diffusion of a solute occurs down? up? its concentration gradient.

___

15 The net diffusion of any substance—solute or solvent—occurs down its own concentration gradient. Consequently, net diffusion tends to produce an equal? unequal? distribution or concentration of any substance.

___

16 Applying the first law of diffusion, the direction of net diffusion of water is down the .... concentration gradient.

___

17 Suppose a nonliving membrane permeable to Na ions, Cl ions, and water separates a 5% sodium chloride solution from a 1% sodium chloride solution.

a Net diffusion of water would occur from the 1%? 5%? solution into the other solution.

___

b Net diffusion of sodium Na ions and Cl ions would occur from the 1%? 5%? solution into the other solution.

___

18 In the situation described in question 17, the concentrations of the two solutions would eventually become equal because net diffusion would:

a Add solute? water? to the 5% solution.

___

b And at the same time, add solute? water? to the 1% solution.

___

■ Movement of substances through cell membranes—cont'd

Procedure B—Experiment demonstrating osmosis

Equipment

1 Beakers
2 Cellophane (porous type) or parchment paper
3 Narrow lumen glass tubing—2 feet or more long
4 Medicine dropper or pipette
5 Osmometer, commercial type
6 Dried seedless raisins
7 Rubber bands
8 Stand and clamps
9 String
10 Distilled water
11 10% and 25% cane sugar solutions

Problems

1 What is osmosis?
2 What is osmotic pressure?
3 In which direction does net osmosis occur?

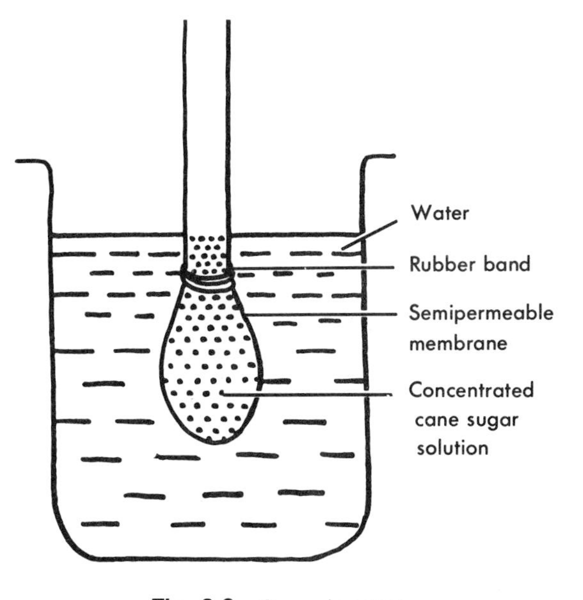

Fig. 2-2 Osmosis setup.

Collection of data

1 If a commercial osmometer is not available, make one as follows:
  a Cut a round-shaped piece of cellophane or parchment paper 6 inches in diameter.
  b Fit this around one end of a long, narrow glass tubing, holding it in place with string (see Fig. 2-2).
  c Wind a cut rubber band around, stretching it as you wind.
  d Tie securely.
  e Submerge the bag in water. Blow through to test for leaks.
2 Clamp the tube to the stand and pour 10% cane sugar solution into the osmometer until the bag is full and the solution rises a short distance in the tube. Cane sugar is nondiffusible through cellophane and parchment paper.
3 Submerge the bag in a beaker of distilled water until the level of the solution in the tube and the level of the water in the beaker are the same. At intervals during the rest of the laboratory period, observe the level of the solution in the tube and the level of the water in the beaker. Observe again the next day. Results?

____

____

____

____

____

Conclusions

1 The changes observed in the level of the fluid in the tube and the level of the water in the beaker indicate that net movement of water (net osmosis, in this case) has occurred into? out of? the 10% sugar solution.

____

2 According to one definition, osmosis is the movement of water through a membrane that, like the one used in this experiment, is? is not? freely permeable to all solute particles present.

---

3 Suppose that a membrane freely permeable to sucrose, salt, and water separates a 5% sucrose solution from a 1% salt solution. Applying the conclusion you drew in question 2, water will pass through this membrane in both directions by the process of diffusion? osmosis?

---

4 If this membrane were freely permeable to salt and water but impermeable to sucrose, water would pass through it by the process of diffusion? osmosis?

---

5 The results of the osmosis experiment you performed illustrate a principle about osmosis that you ought to remember. It is this: although osmosis occurs in both directions through a membrane, net osmosis always takes place into the solution having the higher? lower? concentration of the solute that cannot diffuse freely through the given membrane.

---

6 Osmotic pressure is pressure that develops in a fluid because of net osmosis into it. In the preceding experiment, osmotic pressure develops in the distilled water? sugar solution?

Self-test

1 Can solutes move through membranes by diffusion?

____

2 Can solutes move through membranes by osmosis?

____

3 By what physical process does water move through a membrane that is freely permeable to all the solutes present on both sides of the membrane—by diffusion or by osmosis?

____

4 Down what kind of a gradient does net diffusion of solutes take place?

____

5 Down what kind of a gradient does net osmosis take place?

____

Situation: A membrane permeable to water and impermeable to albumin separates a 30% albumin solution from a 10% albumin solution.

6 Because this membrane is not permeable to any solutes present, it is more precise to say that water diffuses? osmoses? through the membrane than to use the other term.

____

7 Water will move into the 30% solution from the 10% solution and also? but not? in the opposite direction.

____

8 Water is more concentrated in the 10%? 30%? solution because solute is less? more? concentrated in this solution.

____

____

9 Net osmosis always occurs down? up? a water concentration gradient.

____

10 Net osmosis always occurs down? up? a solute concentration gradient.

____

11 In other words, net osmosis occurs into the more? less? concentrated solution.

____

12 For a period of time, more water will move into the 10%? 30%? solution than in the opposite direction.

____

13 After a period of time, the two solutions separated by this membrane each have a 20% albumin concentration. Osmosis still? no longer? goes on between them.

____

14 After equilibration of these two solutions occurs, net osmosis still? no longer? goes on between them.

____

15 Summarizing, after two solutions have equilibrated, osmosis continues in both directions but net osmosis continues in one direction only? no longer occurs?

____

UNIT TWO

THE ERECT AND MOVING BODY

3 The skeletal system

Bones of human skeleton

Procedure A—Identification of bones

Equipment

1. Articulated human skeleton
2. Wall charts of skeleton as a whole and of various parts
3. Illustrations of the following:
   - a Skull—several views
   - b Vertebral column
   - c Thorax
   - d Upper extremities, including shoulder girdle
   - e Lower extremities, including hip girdle
   - f Auditory ossicles
4. Anthony, C. P., and Kolthoff, N. J.: Textbook of anatomy and physiology, ed. 9, St. Louis, 1975, The C. V. Mosby Co., or other textbooks

Problems

1. What bones compose the skull, spinal column, and chest (axial division of the skeleton)?
2. What bones compose the upper and lower extremities (appendicular skeleton)?

Collection of data

1. Identify each of the bones listed below as well as the hyoid bone by:
   - a Finding each bone in an illustration and reading a description of it in your textbook
   - b Locating each bone on the skeleton
   - c Locating each bone as nearly as possible on your own body

Bones of skull

Frontal
Parietal
Temporal
Occipital
Sphenoid
Ethmoid
Nasal
Maxillary
Zygomatic
Mandible
Lacrimal
Palatine
Inferior conchae
Vomer
Malleus
Incus
Stapes

Vertebral column

Cervical vertebrae
Thoracic vertebrae
Lumbar vertebrae
Sacrum
Coccyx

Rib cage

True ribs
False ribs
Sternum

Bones of upper extremities

Clavicle
Scapula
Humerus
Radius
Ulna
Carpals
Metacarpals
Phalanges

Bones of lower extremities

Pelvic bone (os coxae)
Femur
Patella
Tibia
Fibula
Tarsals
Metatarsals
Phalanges

Conclusions

1 Identify the bones shown in Figs. 3-1 to 3-5 by writing the appropriate bone name after each label.

2 What is the name of the U-shaped bone in the neck?

____

3 What are the names of the three small bones in the middle ear?

____

____

____

4 What is the name of the bone that helps form the floor of the cranium and is shaped somewhat like a bat with its wings outstretched?

____

5 What is the name of the bone that makes up a small part of the anterior portion of the cranial floor and the upper part of the nasal septum?

____

6 What bones enter into the formation of the longitudinal arch of the foot?

____

7 What bones form the transverse arch of the foot?

____

8 By what other name is the transverse arch known?

____

9 How are the foot bones held in their arched positions?

____

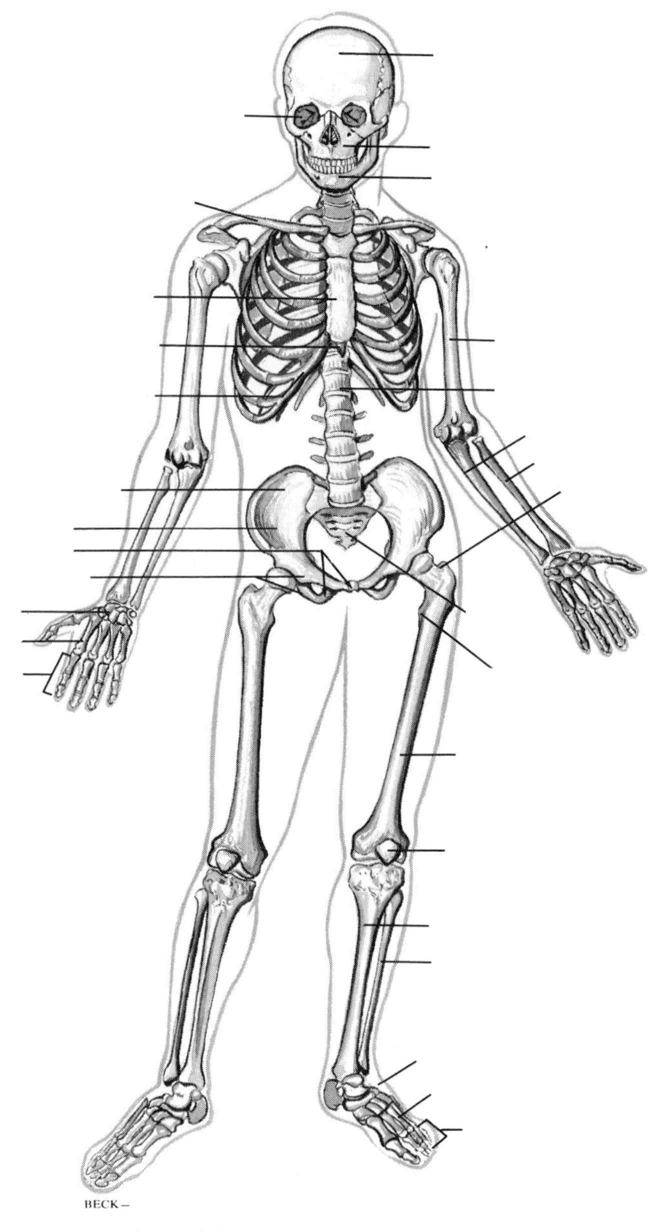

Fig. 3-1 Skeleton—anterior view.

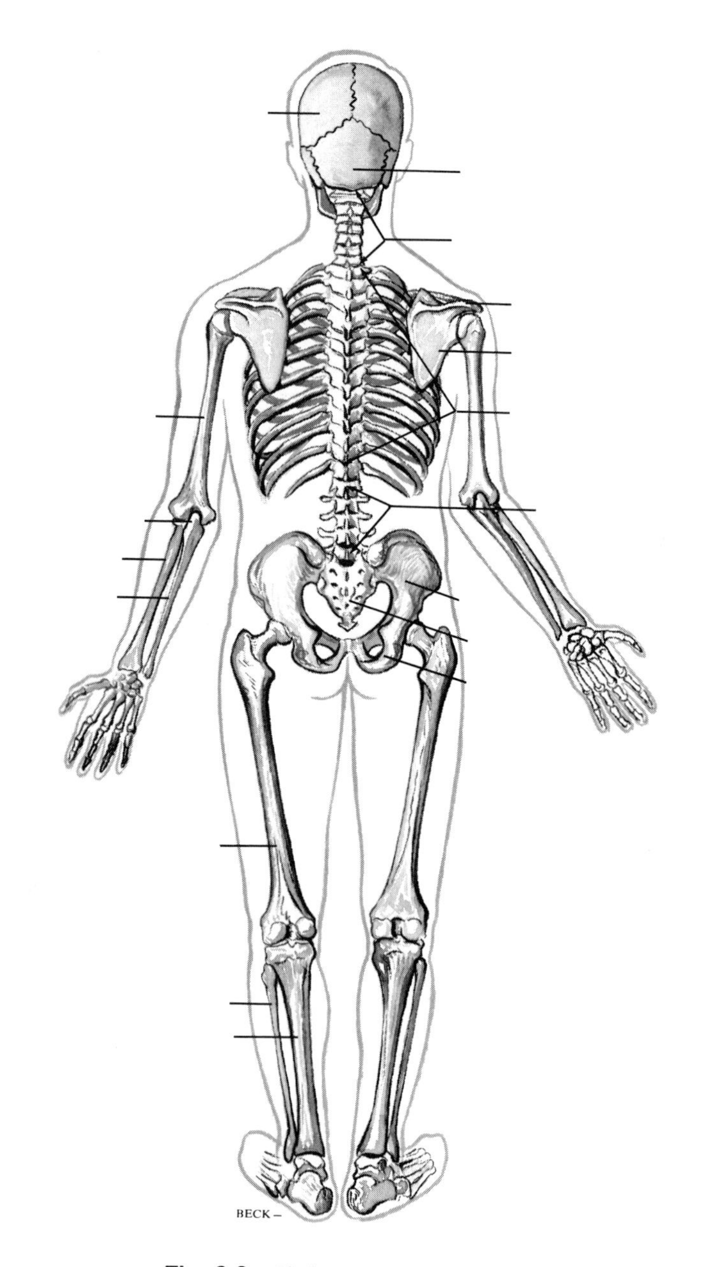

Fig. 3-2 Skeleton—posterior view.

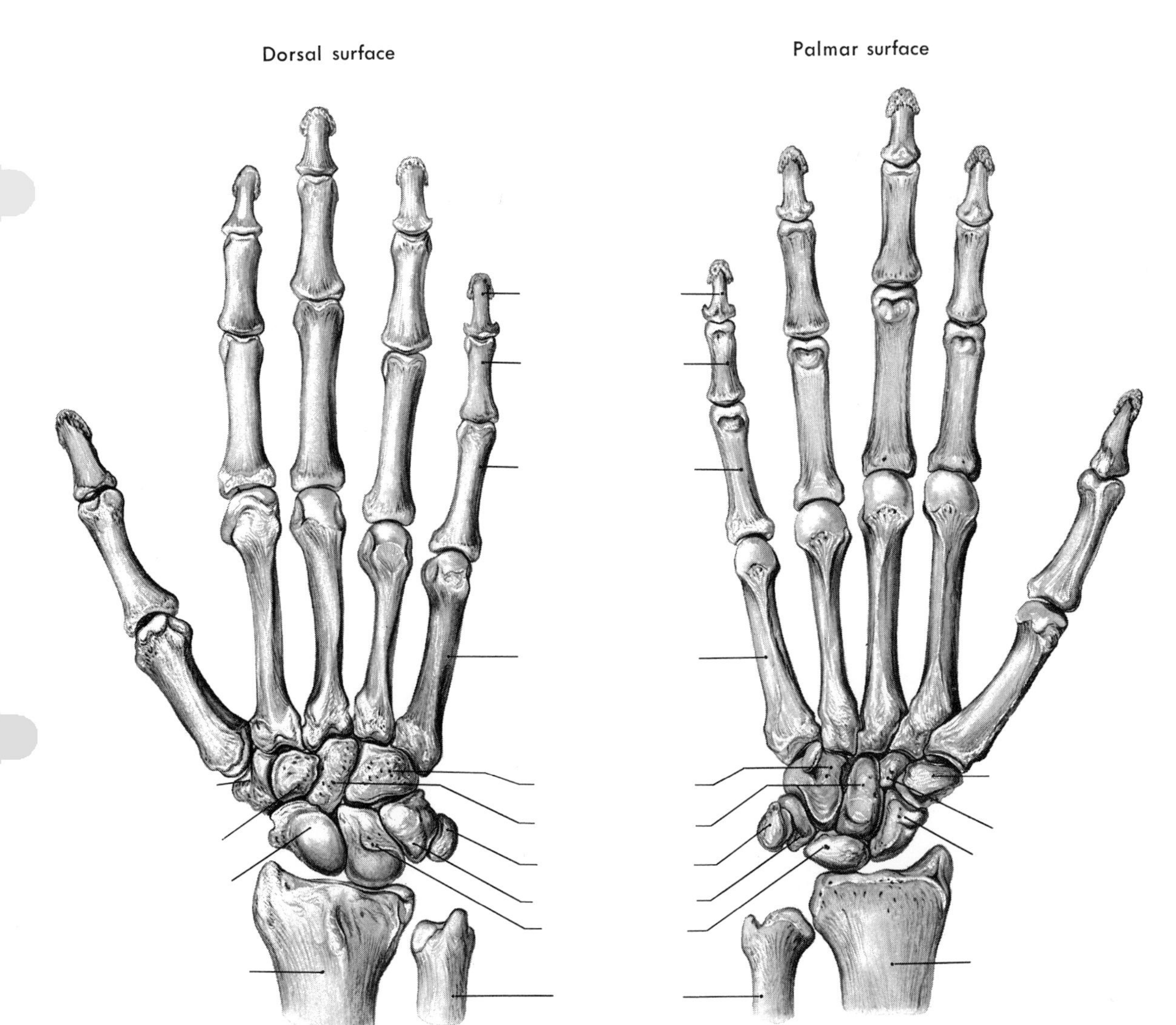

Figs. 3-3 and 3-4 Right hand and wrist. Identify each of the bones listed below by (1) reading a description of the bone in your textbook and finding it in an illustration; (2) locating each bone on the skeleton; (3) writing the name of each bone opposite the label line which points to it:

Carpals:
- Capitate
- Hamate
- Lunate
- Pisiform
- Triquetrum
- Scaphoid
- Trapezium
- Trapezoid

Metacarpals

Phalanges:
- Distal
- Middle
- Proximal

Radius

Ulna

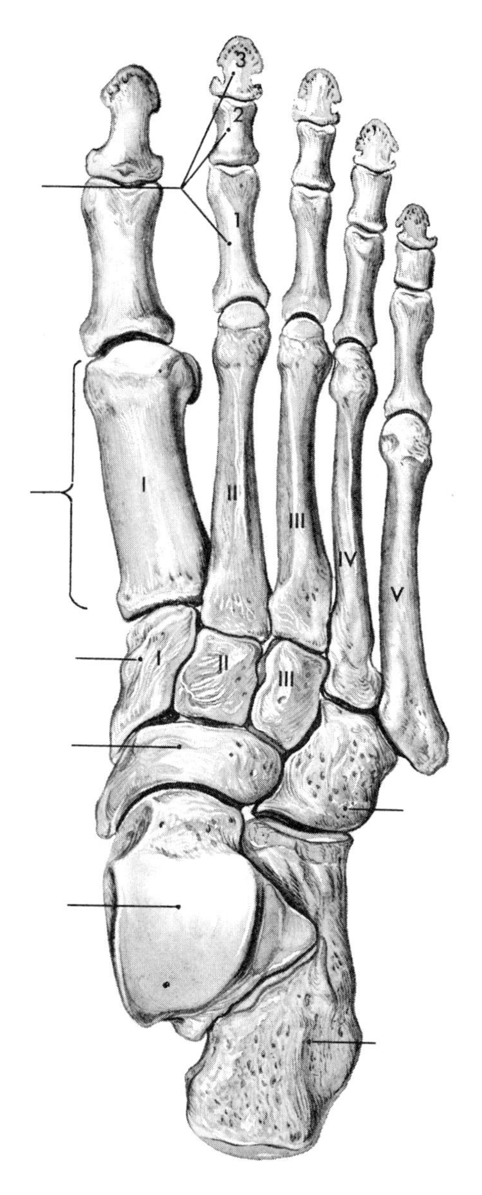

Fig. 3-5 Foot. Identify each of the bones listed below by (1) reading a description of the bone in your textbook and finding it in an illustration; (2) locating each bone on the skeleton; (3) writing the name of each bone opposite the label line which points to it:

Calcaneus	Navicular
Cuboid	Phalanges
Cuneiforms	Talus
Metatarsals	Tarsals (bracket these)

■ Bones of human skeleton—cont'd

Procedure B—Identification of bone markings

Equipment

1. Skeleton
2. Disarticulable skull
3. Infant skull
4. Separate vertebrae
   a. Atlas
   b. Axis
   c. Several thoracic vertebrae
5. Wall charts of skeleton and various parts
6. Anthony, C. P., and Kolthoff, N. J.: Textbook of anatomy and physiology, ed. 9, St. Louis, 1975, The C. V. Mosby Co., or another textbook on anatomy
7. Anatomical atlas

Problem

What are some of the main markings on individual bones?

Conclusions

1. Follow the directions included with Figs. 3-6 to 3-22.
2. Examine the ribs and sternum on the skeleton and answer the following questions:
   a. With what bones do the ribs articulate posteriorly?

      ____

   b. The costal cartilages of which ribs attach directly to the sternum?

      ____

   c. Which ribs are floating—i.e., completely detached from the sternum?

      ____

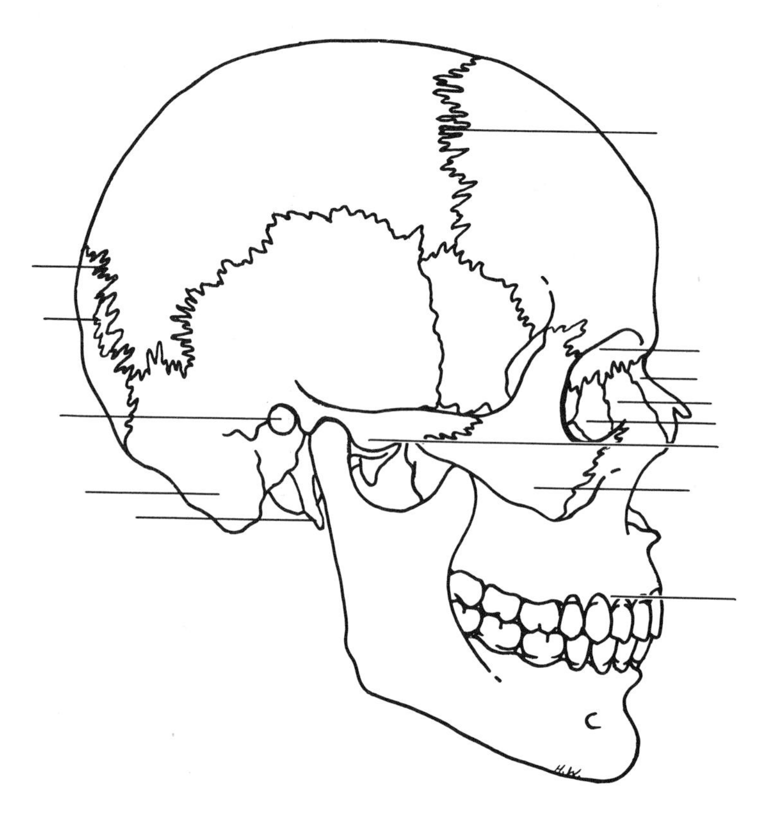

Fig. 3-6 Skull (lateral view). Identify each of the following bone markings by (1) reading a description of the marking in your textbook and finding it in an illustration; (2) locating each marking on the skull; (3) locating each marking, if possible, on your body; (4) writing the name of each marking opposite the appropriate label line:

Alveolar process
Coronal suture
External auditory meatus
Lambdoidal suture
Mastoid process
Styloid process of temporal bone
Supraorbital margin
Wormian bone
Zygomatic arch

Write the name of each of the following bones on the appropriate label line:

Lacrimal bone
Zygomatic bone
Maxilla
Nasal bone

Label and color the following:

Frontal bone . . . . brown
Parietal bone . . . . green
Temporal bone . . . . . . red
Occipital bone . . . . . . brown
Ethmoid bone . . . . blue
Maxillary bone . . . . . . green
Malar bone . . . . lead pencil
Mandible . . . . blue
Nasal bone . . . . red
Lacrimal bone . . . . . orange
Sphenoid bone . . . . yellow

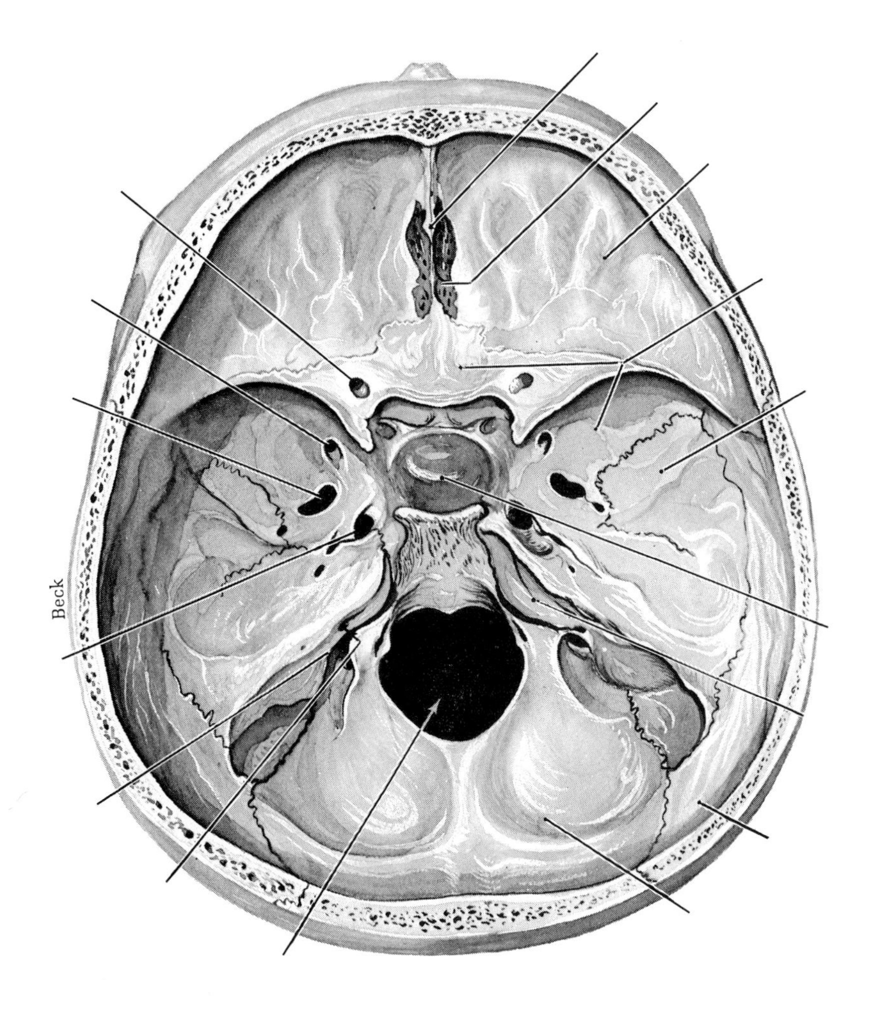

Fig. 3-7 Floor of cranial cavity. Identify each of the following bone markings by (1) reading a description of the marking in your textbook and finding it in an illustration; (2) locating each marking on a skull; (3) writing the name of each marking opposite the appropriate label line, leaving other label lines blank:

Cribriform plate of ethmoid bone
Crista galli of ethmoid bone
Foramen magnum
Frontal bone
Occipital bone
Sella turcica of sphenoid bone

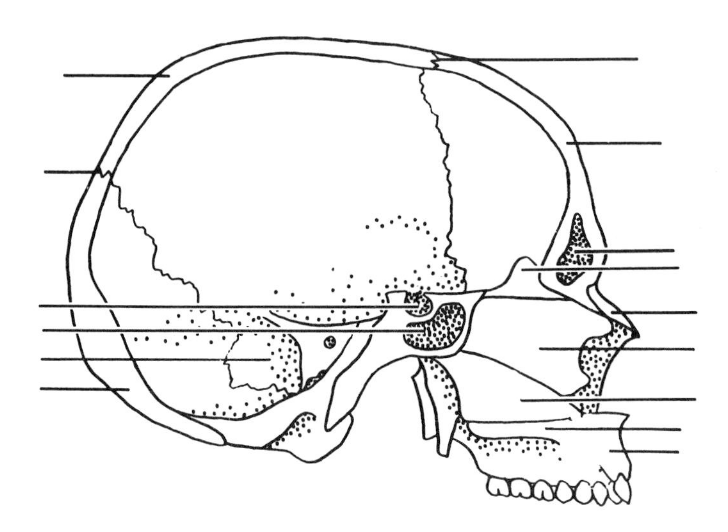

Fig. 3-8 Skull (longitudinal section, left half). Identify each of the following bone markings by (1) reading a description of the marking in your textbook and finding it an an illustration; (2) locating each marking on a skull; (3) writing the name of each marking opposite the appropriate label line:

Coronal suture
Crista galli
Frontal bone
Frontal sinus
Hard palate of maxilla
Lambdoidal suture
Nasal bone
Occipital bone
Parietal bone
Perpendicular plate of ethmoid bone
Sella turcica
Sphenoid sinus
Temporal bone
Vomer bone

Viewed from behind

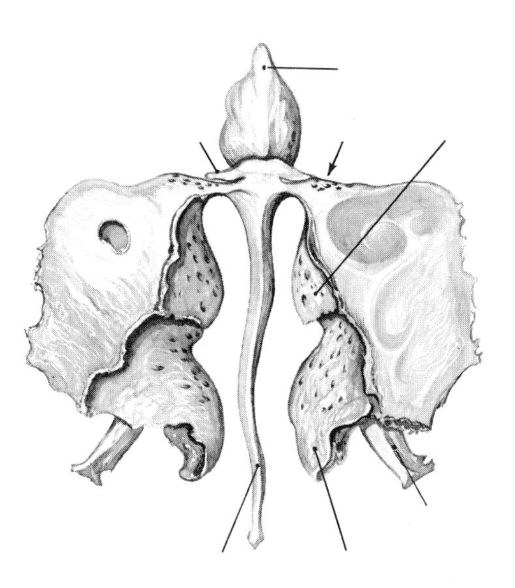

Fig. 3-9

Viewed from above

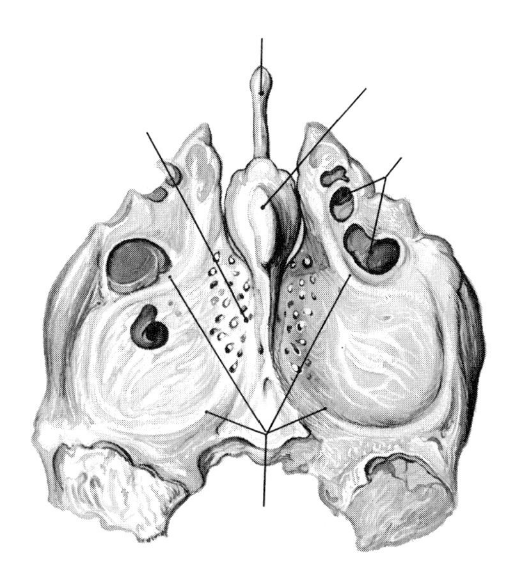

Fig. 3-10

Figs. 3-9 and 3-10 Ethmoid. Identify the following parts of the ethmoid bone by (1) reading a description of the part in your textbook and finding it in illustrations; (2) locating each part on the skull or a disarticulated ethmoid bone; (3) writing the name of each part opposite the appropriate label line, leaving the other label lines blank:

Cribriform (horizontal) plate
Crista galli
Middle concha (turbinate)
Perpendicular plate
Superior concha (turbinate)

Draw a label line pointing to the nasal cavity.

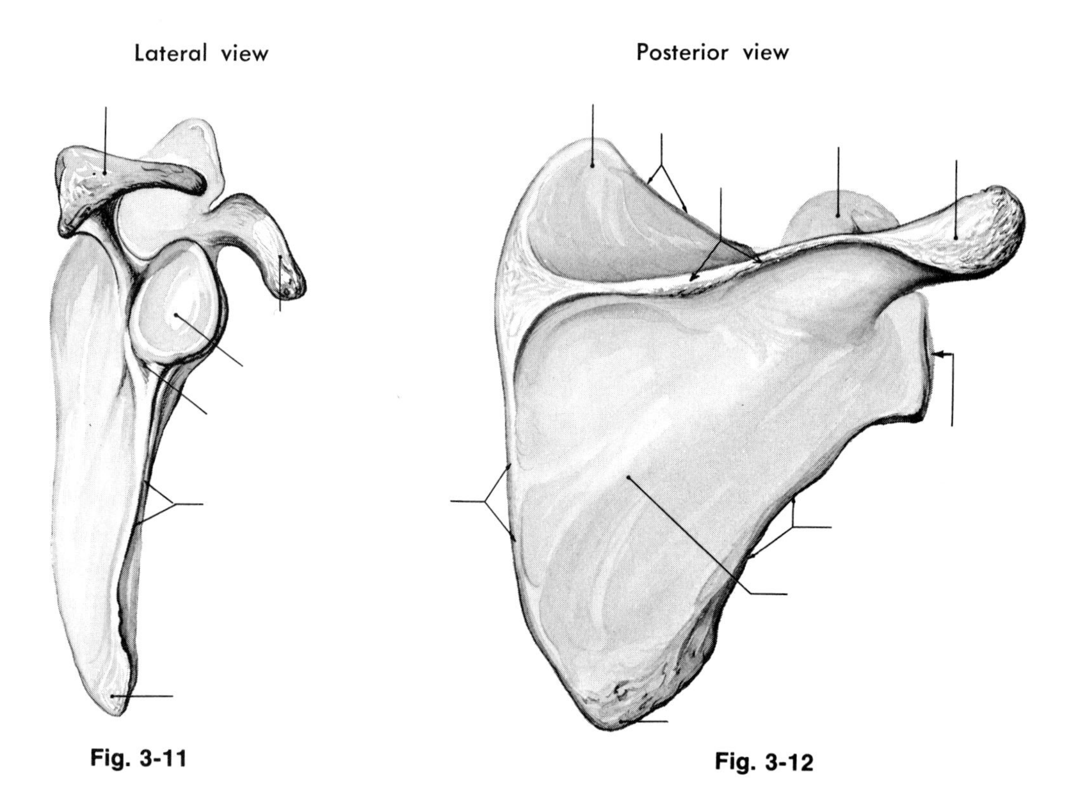

Figs. 3-11 and 3-12 Scapula. Identify each of the following bone markings by (1) reading a description of the marking in your textbook and finding it in an illustration; (2) locating each marking on the skeleton; (3) locating the marking, if possible, on your own body; (4) writing the name of each marking opposite the label line which points to it, leaving the other label lines blank:

Acromion process
Axillary border
Coracoid process
Glenoid cavity
Vertebral border

Anterior view

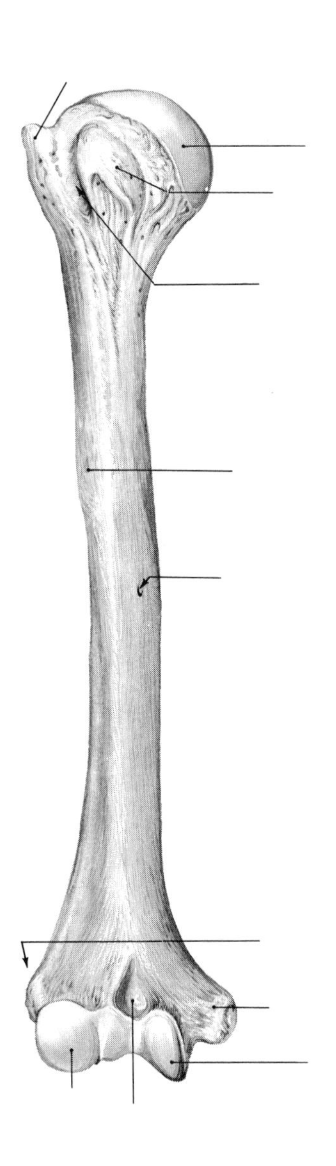

Fig. 3-13

Posterior view

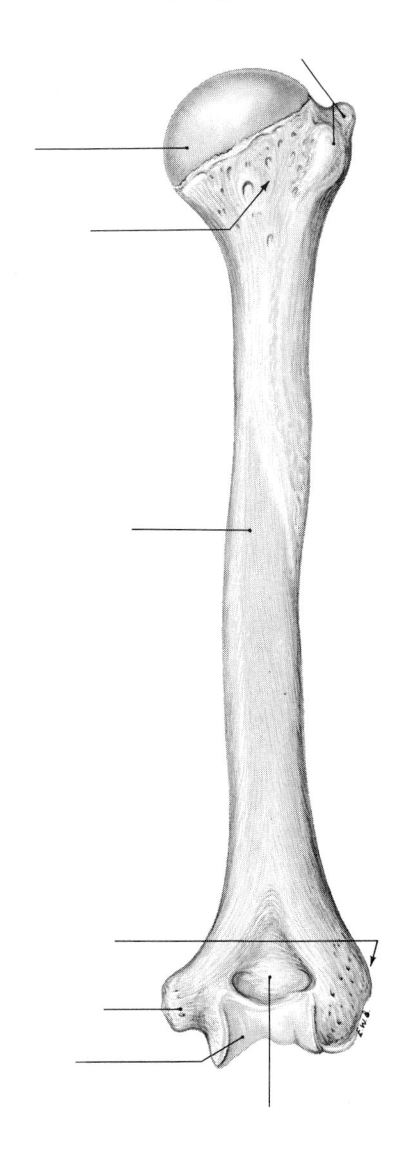

Fig. 3-14

Figs. 3-13 and 3-14 Humerus. Identify each of the following bone markings by (1) reading a description of the marking in your textbook and finding it in an illustration; (2) locating each marking on the skeleton or disarticulated humerus; (3) writing the name of each marking opposite the label line which points to it, leaving the other label lines blank:

Capitulum
Coronoid fossa
Deltoid tuberosity
Head
Lateral epicondyle
Medial epicondyle
Olecranon fossa
Surgical neck
Trochlea

Anterior surfaces

Posterior surfaces

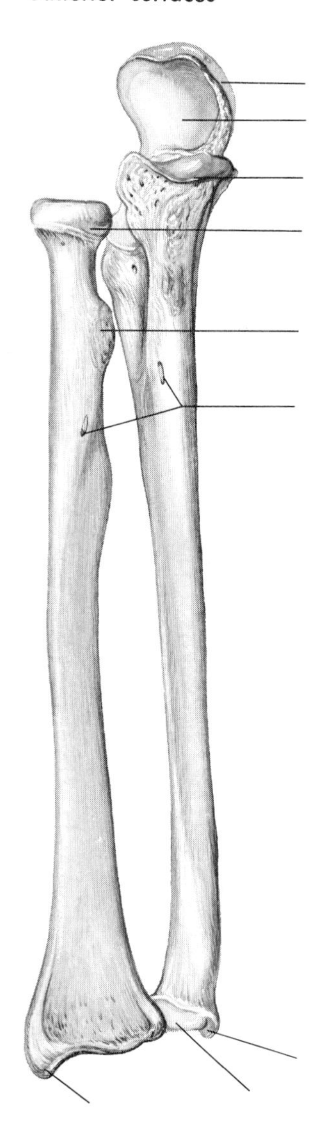

Fig. 3-15

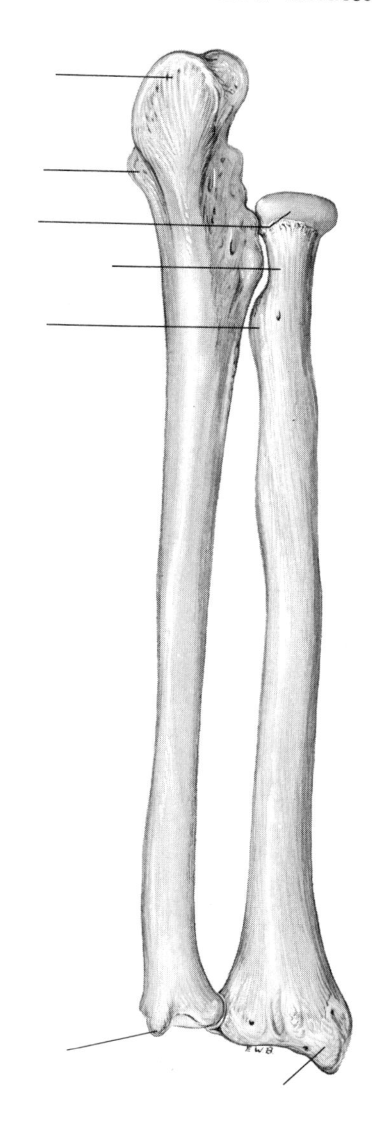

Fig. 3-16

Figs. 3-15 and 3-16 Radius and ulna. Identify each of the following bone markings by (1) reading a description of the marking in your textbook and finding it in an illustration; (2) locating each marking on the skeleton or disarticulated radius and ulna; (3) writing the name of each marking opposite the label line which points to it, leaving the other label lines blank:

Coronoid process
Deltoid tuberosity
Head of radius
Head of ulna
Olecranon process
Radius
Semilunar notch
Styloid process of radius
Styloid process of ulna
Ulna

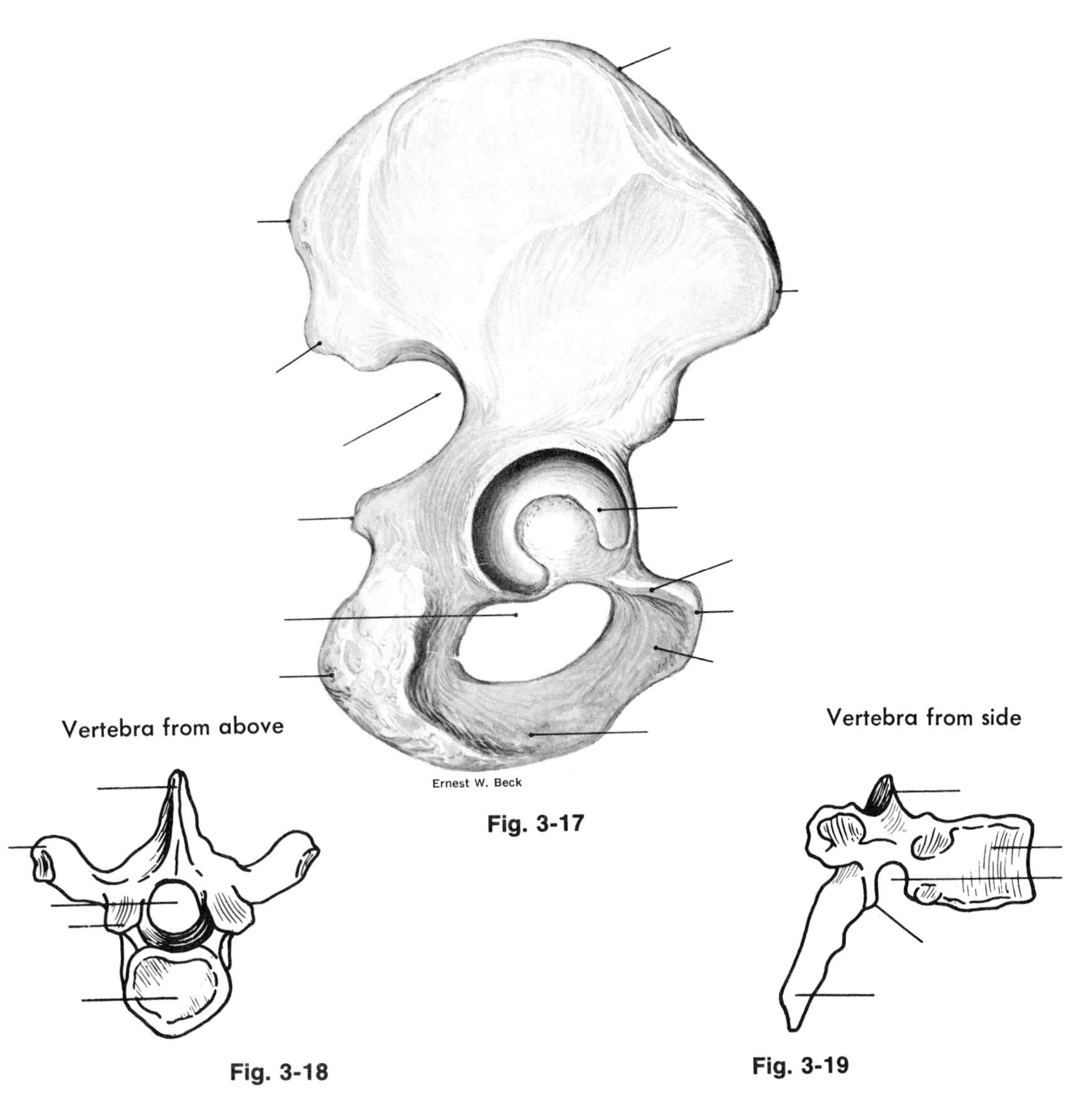

Fig. 3-17

Fig. 3-18

Fig. 3-19

Figs. 3-17 to 3-19 Hip bone and vertebrae. Identify each of the following bone markings by (1) reading a description of the marking in your textbook and finding it in an illustration; (2) locating each marking on the skeleton or disarticulated bones; (3) writing the name of each marking opposite the label line which points to it, leaving other label lines blank:

Acetabulum
Anterior superior spine of ilium
Body of vertebra
Iliac crest
Inferior articulating process
Intervertebral notch
Ischial tuberosity
Pubic crest
Obturator foramen
Spinal foramen
Spine of ischium
Spinous process
Superior articulating process
Transverse process

Color the following:

Ilium .... brown
Pubic bone .... purple
Ischium .... yellow

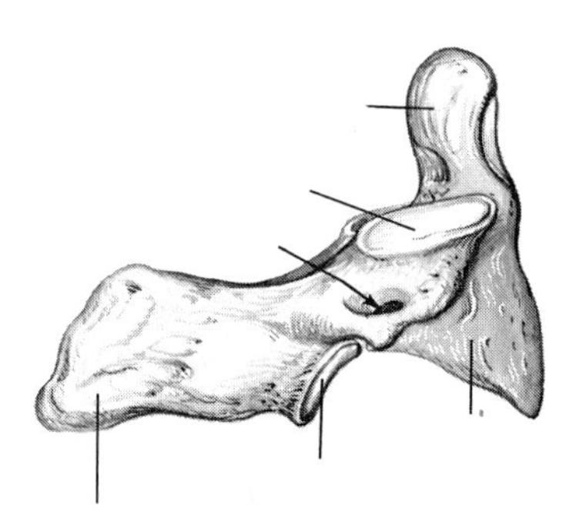

Fig. 3-20 Second cervical vertebrae (axis or epistropheus). Label the odontoid process (dens), leaving the other label lines blank.

■ ■ ■

Make a sketch to show the four curves of the adult spine. Label each curve and indicate how many vertebrae enter into its formation and whether the curve is primary or secondary.

Femur, anterior surface

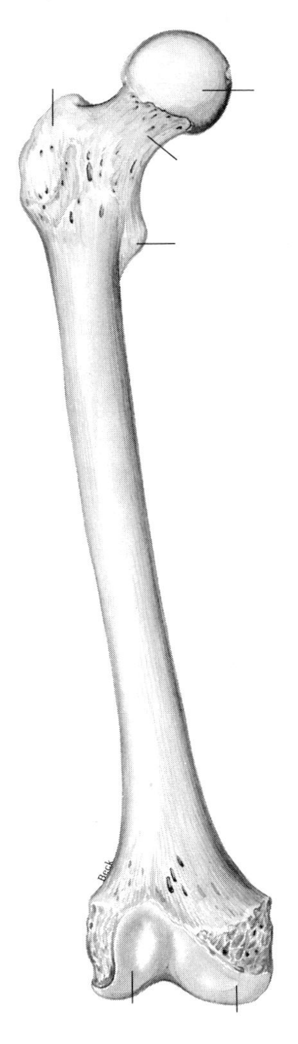

Fig. 3-21

Tibia and fibula, anterior surface

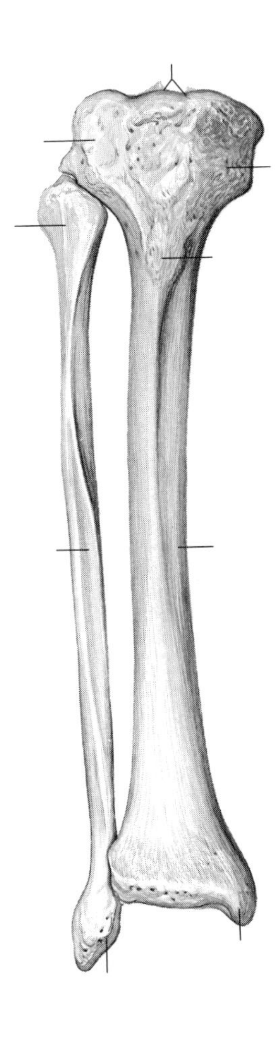

Fig. 3-22

Figs. 3-21 and 3-22 Femur and tibia and fibula. Identify each of the following bone markings by (1) reading a description of the marking in your textbook and finding it in an illustration; (2) locating each marking on the skeleton or disarticulated femur and tibia and fibula; (3) writing the name of each marking opposite the label line which points to it, leaving the other label lines blank:

Fibula	Lesser trochanter
Greater trochanter	Medial condyle
Head of femur	Medial malleolus
Head of fibula	Neck of femur
Lateral condyle	Tibia
Lateral malleolus	Tibial tuberosity

Bones of human skeleton—cont'd

Procedure C—Examination of gross structure of long bone

Equipment

1 Beef bone sawed in half longitudinally
2 Beef bone sawed in half horizontally
3 Dissecting instruments

Problem

What gross structural parts compose a long bone?

Conclusion

Follow the directions included with Fig. 3-23.

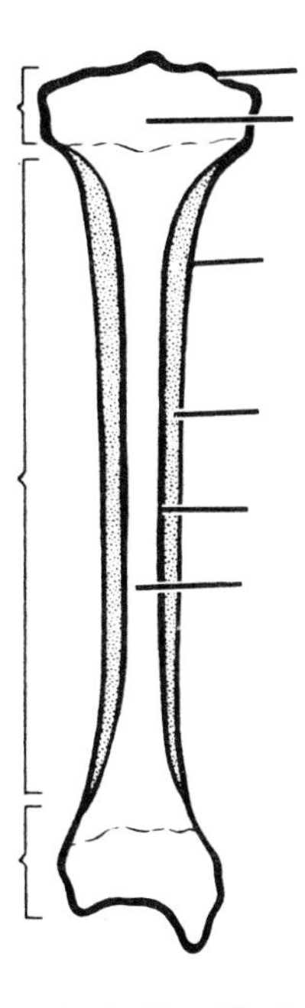

Fig. 3-23 Long bone (longitudinal section). Identify the structures listed below by (1) finding each structure in an illustration and reading a description of it in your textbook; (2) locating each structure on the bone specimens provided; (3) writing the name of the structure to which each label line points:

Articular cartilage	Epiphysis
Compact bone	Medullary (marrow) cavity
Diaphysis	Periosteum
Endosteum	Spongy bone

■ Bones of human skeleton—cont'd

Procedure D—Experiment demonstrating relation between chemical composition and properties of bone

Equipment

1 Two chicken leg bones, soft tissues removed
2 3% acetic acid *or*
3 Vinegar

Problems

1 What properties do the inorganic mineral compounds present in bone give to bone?
2 What properties do the organic constituents of bone give to bone?

Collection of data

1 Soak a chicken leg bone in 3% acetic acid or vinegar for about 24 hours. The inorganic mineral compounds (mainly calcium salts) dissolve out of the bone into the acid. Feel the bone at the end of the soaking period. Try to bend it; to break it. Answer questions 1 and 2 under conclusions.
2 Bake another chicken leg bone in a slow oven (250° to 300° F) for about 2 hours or until it looks well charred. The heat has the effect of removing the organic constituents from the bone. Feel the bone at the end of the baking period. Try to bend it; to break it. Answer questions 3 and 4 under conclusions.

Conclusions

1 Could you bend the bone after it had been soaked in acid?

____

2 Could you break this bone?

____

3 Could you bend the bone after it had been baked?

____

4 Could you break this bone?

____

5 The bone that was flexible and virtually unbreakable was the one that still retained inorganic mineral? organic? compounds.

____

6 The bone that was rigid and easily broken was the one that still retained inorganic mineral? organic? compounds.

____

7 The results of this experiment suggest that bone is a rigid? strong? tissue because it contains organic compounds.

____

8 The results of this experiment suggest that bone is a rigid? strong? tissue because it contains inorganic mineral compounds.

____

■ Joints of human skeleton

Procedure A—Dissection of diarthrotic joint

Equipment

1 Intact beef or lamb joint
2 Beef or lamb joint sawed in half longitudinally
3 Dissecting instruments
4 Dissecting pans
5 Skeleton

Problems

1 What main structural features characterize a diarthrotic joint?
2 What relation exists between joint structure and joint function?

Collection of data

1 Feel the intact joint from the outside. Note whether you can feel the bones distinctly or whether there seems to be some kind of protective padding around them. Answer question 2 under conclusions.
2 On the specimen that has been sawed in half, run your finger over the surface of the membrane that lines the joint cavity and covers the articular surfaces of the bones. Note whether it feels dry or moist, smooth or rough, slippery or sticky. Answer questions 3, 4, and 5 under conclusions.
3 Examine the outer surface of the diaphysis and the articular surface of the epiphysis of one of the specimen bones to see whether they look alike or different. Tap first one surface and then the other with the handle of a scalpel to note whether one seems more resilient and less hard than the other. Answer questions 6, 7, and 8 under conclusions.
4 With a scalpel and tweezers, remove some of the membrane adherent to the diaphysis of one of the bones. Pull on it. Try to push your finger through it. Answer questions 9, 10, and 11 under conclusions.
5 Examine the appearance and test the strength of the structures that hold the two bones together at the joint. Answer questions 12 and 13 under conclusions.
6 Look for cartilage pads in the joint. Feel them. Answer question 14 under conclusions.
7 Note the shapes of the articulating ends of the two adjoining bones in the joint specimen and in several diarthrotic joints on the human skeleton.
8 Examine the sutures in the skull of the human skeleton. Note how their structure makes them immovable joints. Answer question 15 under conclusions.

Conclusions

1 Label Fig. 3-24.
2 From the outside of the joint, the articulating bones felt as if they did not have? did have? a protective padding of soft tissue around them.

____

3 What kind of membrane—mucous, serous, synovial—lines the cavity of a diarthrotic joint and covers the articular surfaces of bones?

____

4 How did synovial membrane feel to you—dry or moist? smooth or rough? slippery or sticky?

____

____

____

5 Based on the way synovial membrane felt to you, do you conclude that it serves the function of facilitating or limiting movement at the joint? Why?

____

6 What kind of tissue covers the articular surfaces of the epiphyses of bones and is itself covered by synovial membrane?

____

7 Based on your observations in step 3

under collection of data, which tissue has more resiliency—bone or cartilage?

____

8 What structure do you conclude functions as a protective cushion over the epiphyses of bones?

____

9 What is the name of the membranous covering adherent to the diaphysis of a bone?

____

10 What kind of tissue composes the periosteum?

____

11 Based on your observation in step 4 under collection of data, would you describe fibrous connective tissue as one of the most delicate tissues or as one of the toughest, strongest tissues in the body?

____

12 What are the strong cords that hold bones together at joints called?

____

13 What kind of tissue composes both ligaments and tendons?

____

14 What structures serve as protective cushions in some joints?

____

15 Complete the following sentence so it states a true principle based on the observations you made in steps 7 and 8 under collection of data: The degree of movement possible at a joint is determined largely by ....

____

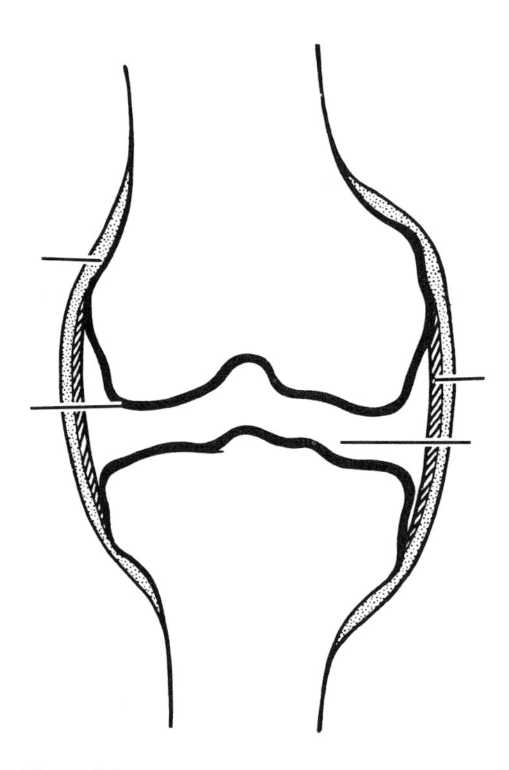

Fig. 3-24 Diarthrotic joint (frontal section).

Joints of human skeleton—cont'd
Procedure B—Study of diarthrotic joint movements

Equipment

1 Articulated skeleton
2 Textbook of anatomy and physiology

Problems

1 What do the following terms mean?

abduction	extension
adduction	flexion
circumduction	rotation

2 Which of the preceding movements are possible at each of the following joints?

shoulder	hip
elbow	knee

3 What structural type of diarthrotic joint allows the greatest range of movement?

Collection of data

1 Consult your textbook for the meanings of the terms listed under problem 1. Answer question 1 under conclusions.
2 Study the *shoulder joint* as follows:
  a Note the shapes of the articulating structures of the shoulder joint—viz., the head of the humerus and the glenoid cavity of the scapula.
  b Starting with the arms in the anatomical position, move them forward and up to shoulder height.
  c Starting with the arms in the anatomical position, move them straight out back as far as possible.
  d Starting with the arms in the anatomical position, move them straight out to the sides.
  e Starting with the arms straight out at the sides, move them straight down to the sides of the body.
  f Starting with your arms straight out at the side, move them as if you were drawing as large a circle as possible with your hands.
  g Starting with your arms straight out at the side, turn the palms upward, then forward, then downward, then backward.
  h Answer question 2 under conclusions (p. 52).
3 Study the *elbow joint* as follows:
  a Examine the shapes of the articulating structures of the elbow joint—viz., trochlea of the humerus and semilunar notch of the ulna.
  b Starting with your right arm straight out from the shoulder, bend your arm so as to touch the fingers of your right hand to your right shoulder.
  c Reverse the movement you have just performed—i.e., start with the final position described in step b and return to the starting position.
  d Answer question 3 under conclusions.
4 Study the *hip joint* as follows:
  a Examine the shapes of the articulating structures at the hip joints—viz., the head of the femur and the acetabulum of the pelvic bone.
  b Starting from the standing position and keeping your knees rigid, move your right leg forward as high as possible.
  c Starting from the standing position and with your knees rigid, move your right leg straight out back.
  d Starting from the standing position, move your right leg straight out to the side.
  e Standing with legs astride, move your right leg medially so that both legs are close together.
  f Answer question 4 under conclusions.
5 Study the *knee joint* as follows:
  a Examine the shapes of the articulating parts of the knee joint—viz., the distal end of the femur and the proximal end of the tibia.
  b Sit down on a chair with both feet flat on the floor. Now kick your right leg

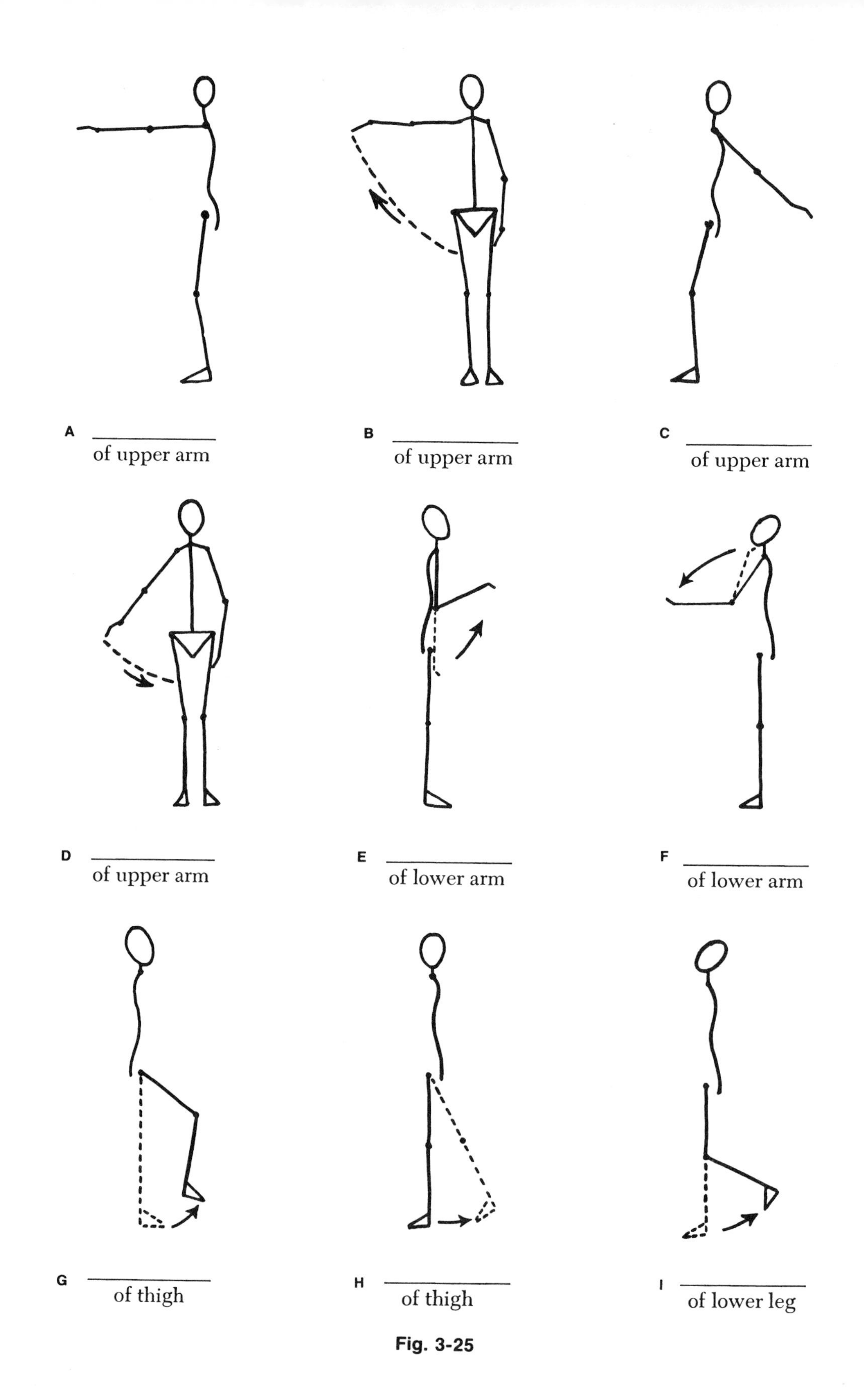
A
of upper arm
B
of upper arm
C
of upper arm
D
of upper arm
E
of lower arm
F
of lower arm
G
of thigh
H
of thigh
I
of lower leg

Fig. 3-25

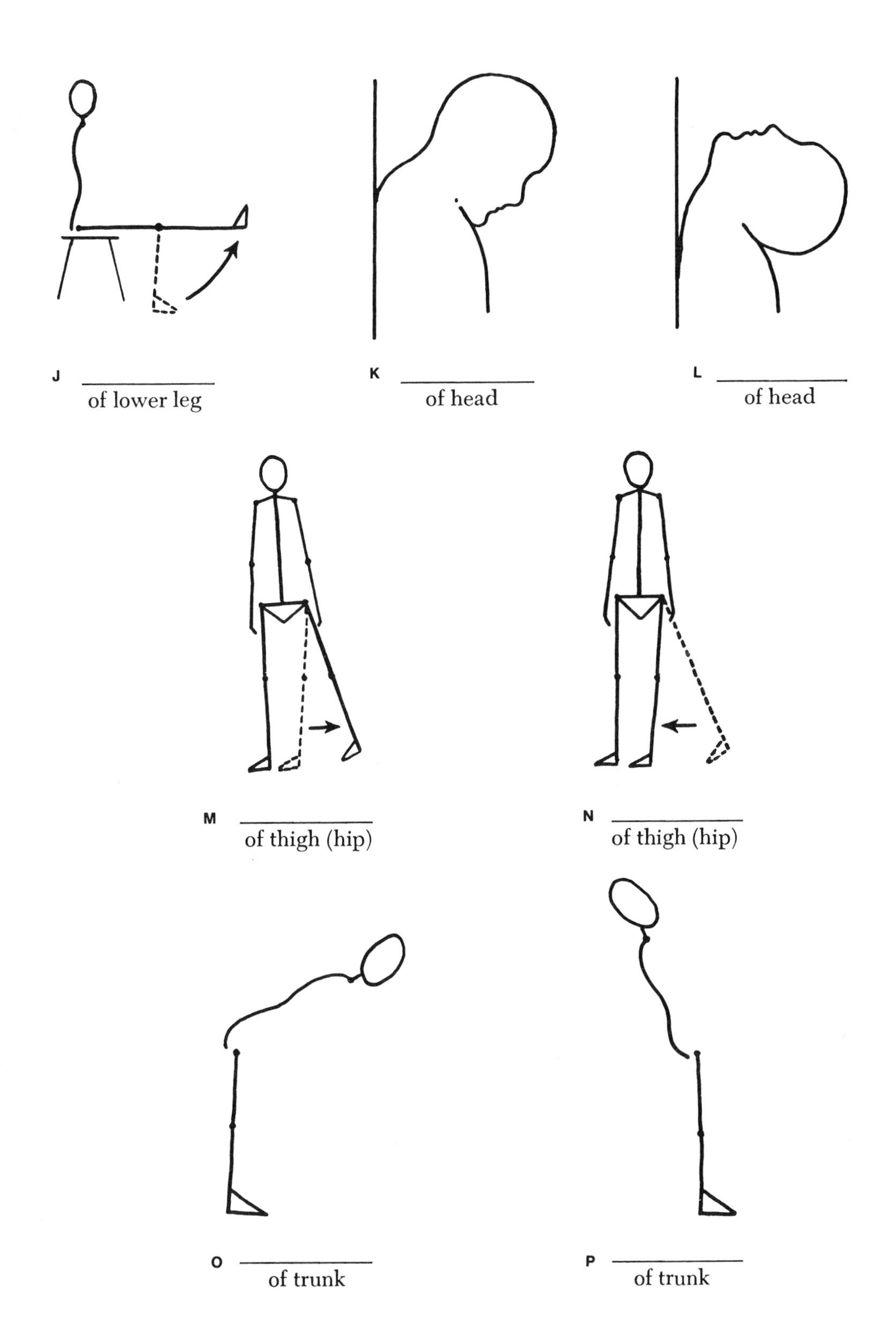

Fig. 3-25, cont'd

forward so that it is straight out in front of you.

c Reverse the movements you have just performed, that is, lower your right leg so that your foot is again flat on the floor.

d Answer question 5 under conclusions.

Conclusions

1 What kind of movement will you be performing if you move a part of your body so as to:

a Decrease the angle between two bones?

____

b Move the part away from the median plane of the body?

____

c Move a bone on its own axis?

____

Answer questions 2 to 7 by writing one of the following terms in the blank spaces provided below each figure in Fig. 3-25:

abduction — extension
adduction — flexion

2 What movements are portrayed by drawings **A** to **D**?

3 What movements are portrayed by drawings **E** and **F**?

4 What movements are portrayed by drawings **G** and **H** and **M** and **N**?

5 What movements are portrayed by drawings **I** and **J**?

6 What movements are portrayed by drawings **K** and **L**?

7 What movements are portrayed by drawings **O** and **P**?

■ Self-test

Various prominent bone markings are listed below. In the space provided, give the name of the bone(s) on which the marking is found.

1 Alveolar processes

____

____

2 Antrum of Highmore

____

3 Acromion process

____

4 Acetabulum

____

5 Anterior superior iliac spine

____

6 Crista galli

____

7 Coracoid process

____

8 Mastoid process

____

9 Foramen magnum

____

10 Cribriform plate

____

11 Mental foramen

____

12 Obturator foramen

____

13 Spinal foramen

____

14 Odontoid process

____

15 Laminae

____

16 Neural arch

____

17 Manubrium

____

18 Glenoid cavity

____

19 Greater trochanter

____

20 Olecranon process

____

21 Symphysis pubis

____

22 Lateral malleolus

____

23 Xiphoid process

____

24 Wormian bones

____

25 Zygomatic process

____

26 The structure of some joints allows all of these movements—flexion, extension, abduction, adduction, circumduction, and rotation. Which of these movements does each of the following joints permit? (If the joint permits all of the movements, write the word all in the answer space; otherwise write the names of the movements possible at the joint.)

a Shoulder joint?

____

b Elbow joint?

____

c Hip joint?

____

d Knee joint?

____

e Joints between the phalanges of the hand?

____

4 The muscular system

Structure, location, and actions of skeletal muscles

Procedure A—Examination of muscles of small mammal

Equipment

1 Freshly killed small mammal (e.g., rat or guinea pig) *or*
2 Embalmed cat or rhesus monkey
3 Dissecting instruments
4 Dissecting board

Problems

1 Where are skeletal muscles placed with relation to bones and joints?
2 What is the functional relationship between skeletal muscles and the skeleton?

Collection of data

1 Remove part or all of the skin from the animal by dissecting away the subcutaneous fascia that attaches the skin to the underlying structures.

2 **a** Locate the large muscle of the chest —viz., the pectoralis major. Note its size and shape and the direction of its fibers.
b Identify the bones to which the pectoralis major attaches. Try to move these bones and determine which moves more easily.
c Pull on the cord that attaches the pectoralis major muscle to a bone.
d Like most muscles, the pectoralis major extends across a joint. Note which one and what type of joint it is. Answer questions 1 to 7 under conclusions.

3 Examine as many more individual muscles as time permits. Note size and shape and direction of fibers, as well as bony attachments. Note also the relation of each muscle to adjacent muscles.

4 Look up the meaning of the term deep fascia. Observe deep fascia on this animal.

5 **a** Cut open the abdominal cavity and remove the viscera. Examine the diaphragm muscle. Note the shape of this important muscle and the kind of tissue that forms its central portion. Identify the bones to which it attaches.
b Take hold of the central tendon of the diaphragm with forceps and pull downward. When the diaphragm contracts, it pulls the central tendon of the diaphragm downward. Note how

the size of the animal's chest cavity changes when you pull down on the central tendon. Contraction of the diaphragm produces the same change in chest cavity size, and this automatically causes inspiration (explained in Chapter 9).

Conclusions

1 To which bones does the pectoralis major muscle attach?

____

____

2 Of the bones to which the pectoralis major attaches, which one seemed to you to be most easily moved? ______

3 What kind of tissue attaches muscle to bone?

____

4 When the tissue that anchors muscle to bone has an elongated shape like a piece of cord, what is this structure called?

____

5 What movement of what part of the animal's body took place as you pulled on the tendon of the pectoralis major muscle?

____

6 Across what joint does the pectoralis major muscle extend?

____

7 What type of joint is the shoulder joint?

____

8 When the diaphragm's fibers contract, does its domelike shape become flatter? more arched?

____

9 As the diaphragm contracts, what change does it produce in the size of the thoracic cavity?

____

Structure, location, and actions of skeletal muscles—cont'd

Procedure B—Examination of muscles that move upper arm

Equipment

Skeleton

Problems

1 What bone necessarily serves as the insertion for all muscles that act as prime movers of the upper arm? The prime movers to be studied are the:
   a Pectoralis major
   b Latissimus dorsi
   c Deltoid
2 At which joint does movement occur whenever the upper arm moves?
3 Where is each muscle located with relation to bones and joints? More specifically:
   a Where is its body located—over what bones and/or joints?
   b Where is its origin—on what bone or bones?
   c Where is its insertion—on what bone or bones?
4 What movement does contraction of each muscle produce?

Collection of data

1 Locate in Figs. 4-1 or 4-2 the pectoralis major, latissimus dorsi, and deltoid muscles.
2 Study the picture of each of these muscles to try to determine:
   a What bones and joints lie under the body of each muscle
   b On what bone or bones it probably has its origin
   c On what bone or bones it probably has its insertion
3 Locate the pectoralis major on the skeleton and as nearly as possible on your own body.
4 Deduce the main movement produced by contraction of the pectoralis major. Apply the following principle in making your deduction. When a muscle contracts, it moves its insertion bone, i.e., it pulls its insertion bone toward its origin bone. Now place your hand over the pectoralis major muscle and perform the movement you have deduced it produces. If you feel the muscle contract as you perform this movement, you will know that your deduction is correct. Answer questions 1, 2, and 3 under conclusions.
5 Repeat steps 2 and 3 for the *latissimus dorsi* muscle. Ask your partner to perform the movement you have reasoned that this muscle produces. Place your hand over it during the movement. Answer question 4 under conclusions.
6 Repeat steps 2, 3, and 4 for the *deltoid* muscle. Answer question 5 under conclusions.

Conclusions

1 What bone necessarily serves as the insertion bone for all muscles that act as prime movers of the upper arm?

____

2 At which joint does movement occur whenever the upper arm moves?

____

3 What movements of the upper arm does contraction of the pectoralis major produce?

____

4 What movements of the upper arm does contraction of the latissimus dorsi produce?

____

5 What movements of the upper arm does contraction of the deltoid produce?

____

6 Judging from your own observations

and deductions, does the following sentence state a true principle? In general, the muscles that move the upper arm are those that form the flesh of the upper arm.

____

7 What muscle serves as the prime abductor of the upper arm?

____

8 What muscles function together as the prime adductors of the upper arm?

____

9 What muscle serves as the prime flexor of the upper arm?

____

10 What muscle serves as the prime extensor of the upper arm?

____

11 Name the antagonist to the pectoralis major muscle.

____

12 Name the antagonists to the deltoid muscle.

____

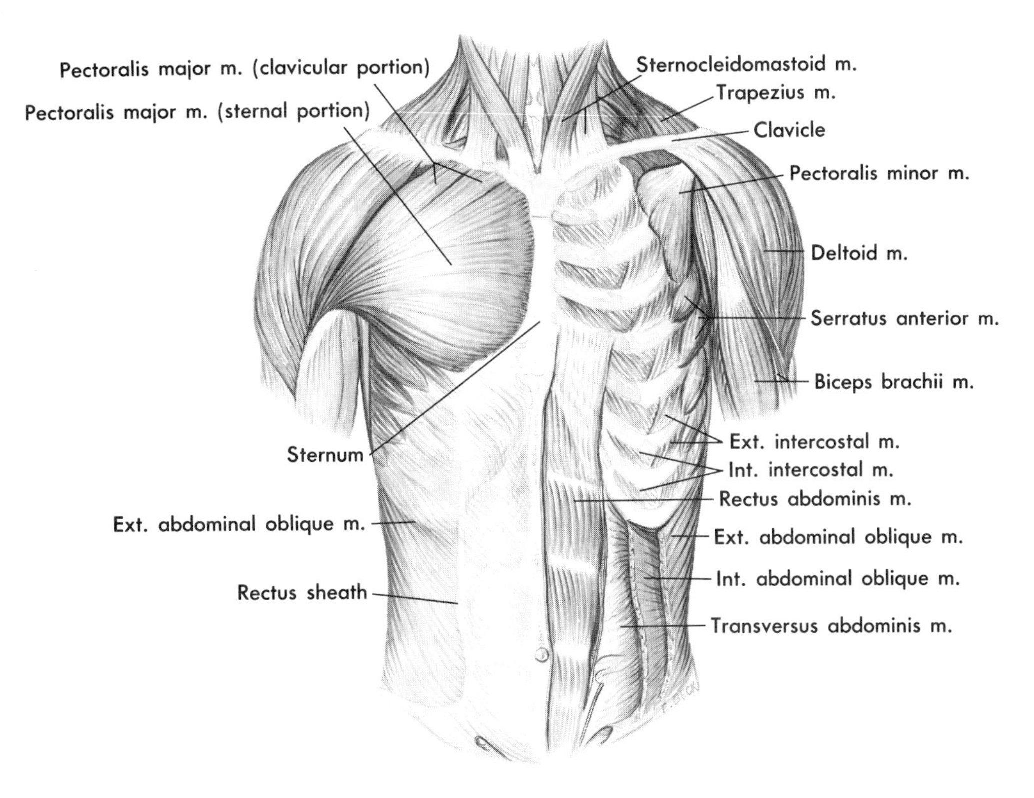

Fig. 4-1 Superficial muscles of anterior surface of trunk.

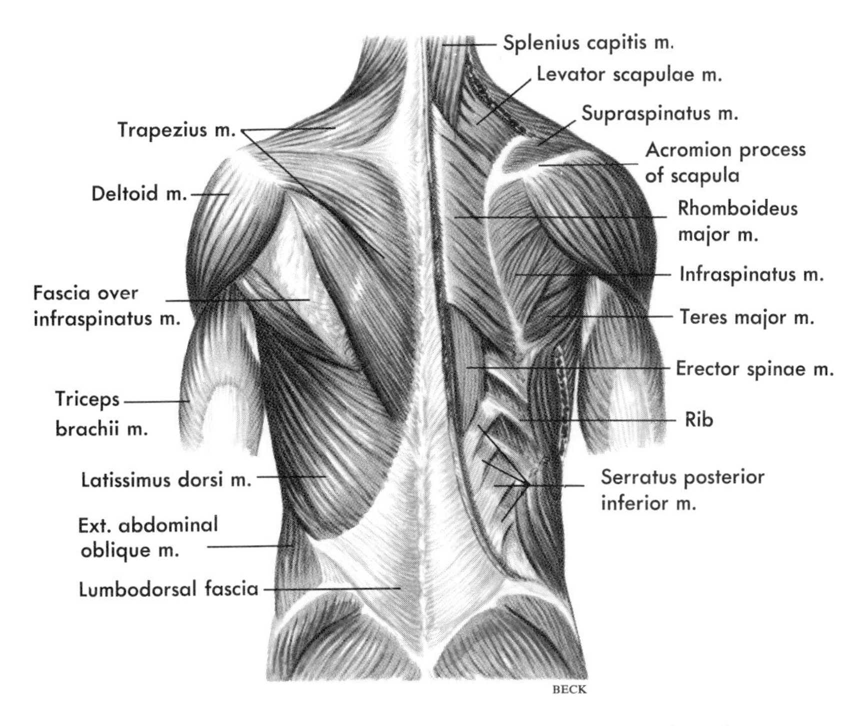

Fig. 4-2 Superficial muscles of posterior surface of trunk.

■ Structure, location, and actions of skeletal muscles—cont'd

Procedure C—Examination of muscles that move lower arm

Equipment

Skeleton

Problems

1 What bones serve as the insertion bones for all muscles that act as prime movers of the lower arm? The prime movers we shall investigate in this procedure are the:
   - a Biceps brachii muscle
   - b Triceps brachii muscle
   - c Brachialis muscle

2 At which joint does movement occur whenever the lower arm moves independently of the upper arm?

3 Where is each muscle located with relation to bones and joints?

4 What movement does contraction of each muscle produce?

Collection of data

1 Study the illustration of the *biceps brachii* (Fig. 4-3) to determine:
   - a What bone lies under the body of the muscle
   - b On what bone or bones it probably has its origin
   - c On what bone or bones it probably has its insertion
   - d What joint or joints the muscle spans

2 Locate the biceps brachii on the skeleton and as nearly as possible on your own body.

3 From the observations you have just made, deduce the main movement produced by contraction of the biceps brachii muscle. Apply the principle you have already learned that when a muscle contracts, it moves its insertion bone toward its origin bone. Place your hand over the biceps brachii muscle and perform the movements you have deduced it produces. If you feel the muscle contract as you perform this movement, you will know your deduction is correct. Answer questions 1, 2, 3, and 4 under conclusions.

4 Repeat steps 1, 2, and 3 for the *triceps brachii* muscle (Fig. 4-4). Answer question 5 under conclusions.

5 Repeat steps 1, 2, and 3 for the *brachialis* muscle (Fig. 4-5). Answer question 6 under conclusions.

Conclusions

1 Identify the origin and insertion bones for the muscles shown in Figs. 4-3 to 4-5 by writing the word origin or insertion on the appropriate label lines.

2 What bone(s) necessarily serve as the insertion bone(s) for all muscles that act as prime movers of the lower arm?

____

3 At which joint does movement occur whenever the lower arm moves independently of the upper arm?

____

4 What movement does contraction of the biceps brachii produce?

____

5 What movement does contraction of the triceps brachii produce?

____

6 What movement does contraction of the brachialis produce?

____

7 Judging from your own observations and deductions, does the following sentence state a true principle? In general, muscles that move the lower arm form the flesh of the upper arm.

____

8 What muscles flex the lower arm?

____

9 What muscle extends the lower arm?

____

10 What muscles act as antagonists to the triceps brachii?

____

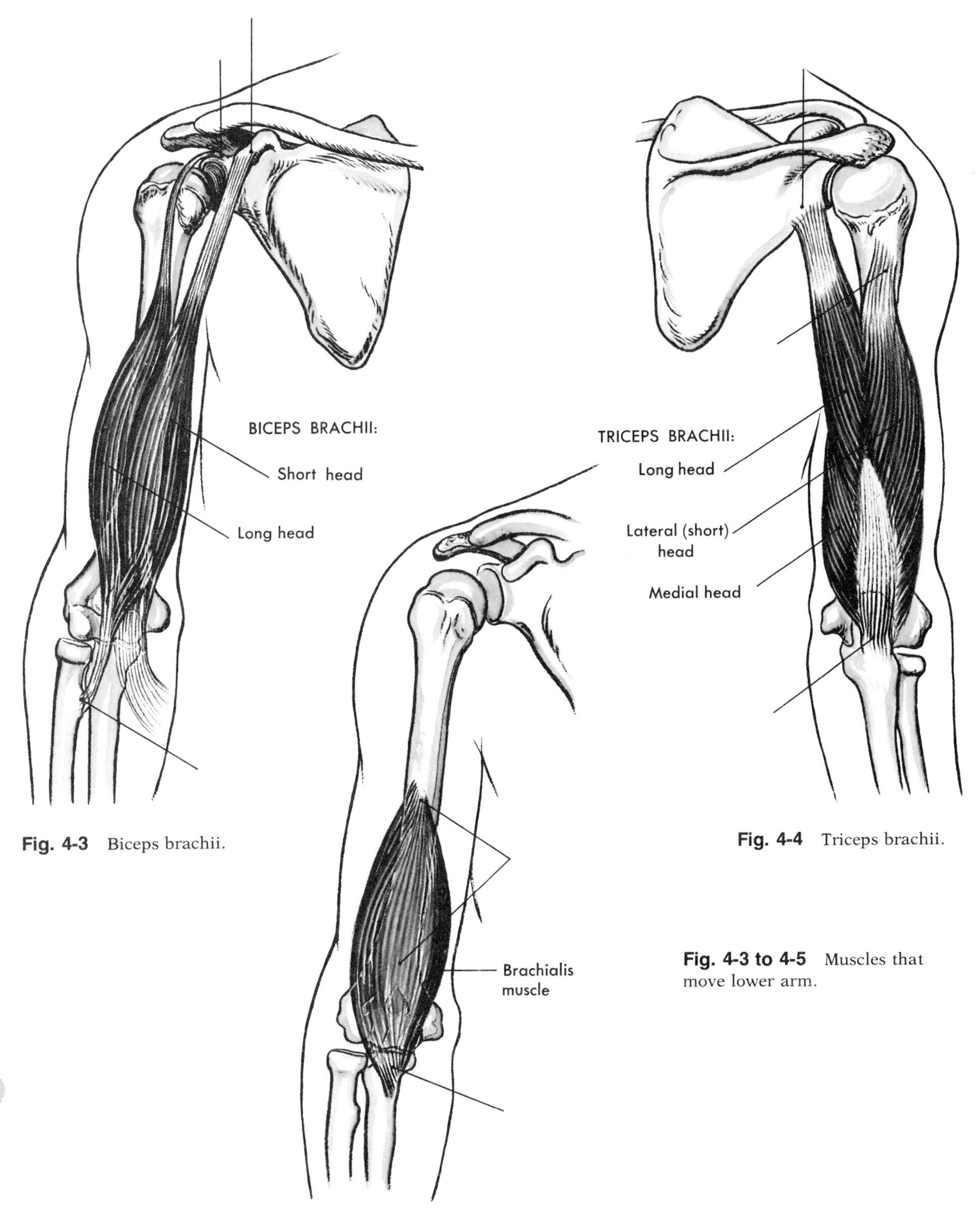

Fig. 4-3 Biceps brachii.

Fig. 4-4 Triceps brachii.

Fig. 4-5 Brachialis.

Fig. 4-3 to 4-5 Muscles that move lower arm.

■ Structure, location, and actions of skeletal muscles—cont'd

Procedure D—Examination of muscles that move thigh

Equipment

Skeleton

Problems

1 What bone serves as the insertion bone for all muscles that act as prime movers of the thigh? The prime movers we shall investigate are the:
   a Iliopsoas
   b Gluteus maximus
   c Gluteus medius and minimus
   d Adductors
2 At which joint does movement occur whenever the thigh moves?
3 Where is each muscle located with relation to bones and joints?
4 What is the main movement produced by the contraction of each muscle?

Collection of data

1 Study the picture of the *iliopsoas* muscle (iliacus and psoas muscles) (Fig. 4-6) to try to determine:
   a What bone or bones lie under the bodies of this compound muscle and what joints it pulls across
   b On what bone or bones it probably has its origin when it functions as a mover of the thigh
   c On what bone it has its insertion when it acts as a mover of the thigh
2 Locate the iliopsoas muscle on the skel-

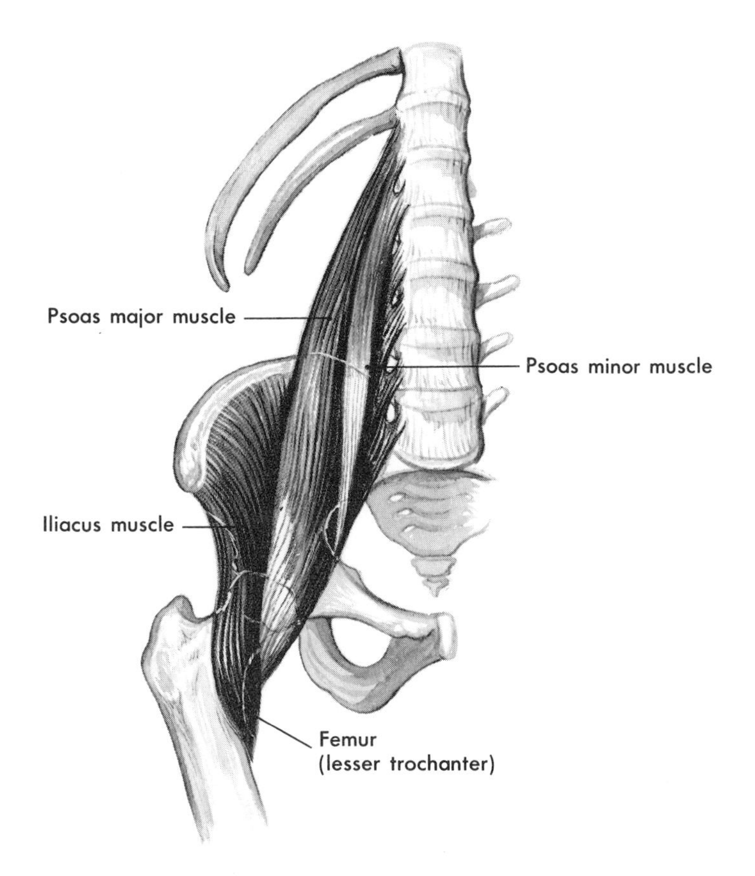

Fig. 4-6 Iliopsoas muscle.

eton and approximately on your own body.

3 Deduce what thigh movement is produced by contraction of the iliopsoas muscle. Answer questions 2, 3, 4, and 5 under conclusions.
4 Repeat steps 1, 2, and 3 for the *gluteus maximus* muscle (Figs. 4-7 and 4-8). Answer question 6 under conclusions.
5 Repeat steps 1, 2, and 3 for the *gluteus medius* muscle (Fig. 4-10) and *gluteus minimus* muscle (Fig. 4-9). Answer question 7 under conclusions.
6 Repeat steps 1, 2, and 3 for the *adductor group* (Fig. 4-11) of muscles. Answer questions 8 through 13.

Conclusions

1 Identify the origin and insertion bones for the muscles shown in Figs. 4-8 to 4-11 by writing the word origin or insertion on the appropriate label lines.
2 What bone necessarily serves as the insertion bone for muscles that act as prime movers of the thigh?

____

3 At which joint does movement occur whenever the thigh moves?

____

4 Does contraction of the iliopsoas flex or extend the thigh? Why?

____

5 Imagine yourself in the standing position. Now imagine that both iliopsoas muscles contract while the lower vertebrae serve as insertions and the femur serves as the origin. What movement do you deduce would result?

____

6 What movement of the thigh does contraction of the gluteus maximus produce?

____

7 What movement of the thigh does contraction of the gluteus medius and minimus produce?

____

8 What movement of what bone does contraction of the adductor group of muscles produce?

____

9 Name the prime abductors of the thigh.

____

10 Name a prime flexor of the thigh.

____

11 Name a prime extensor of the thigh.

____

12 Name the muscles that act as antagonists to the adductors of the thigh.

____

13 Name the muscle that acts as an antagonist to the gluteus maximus.

____

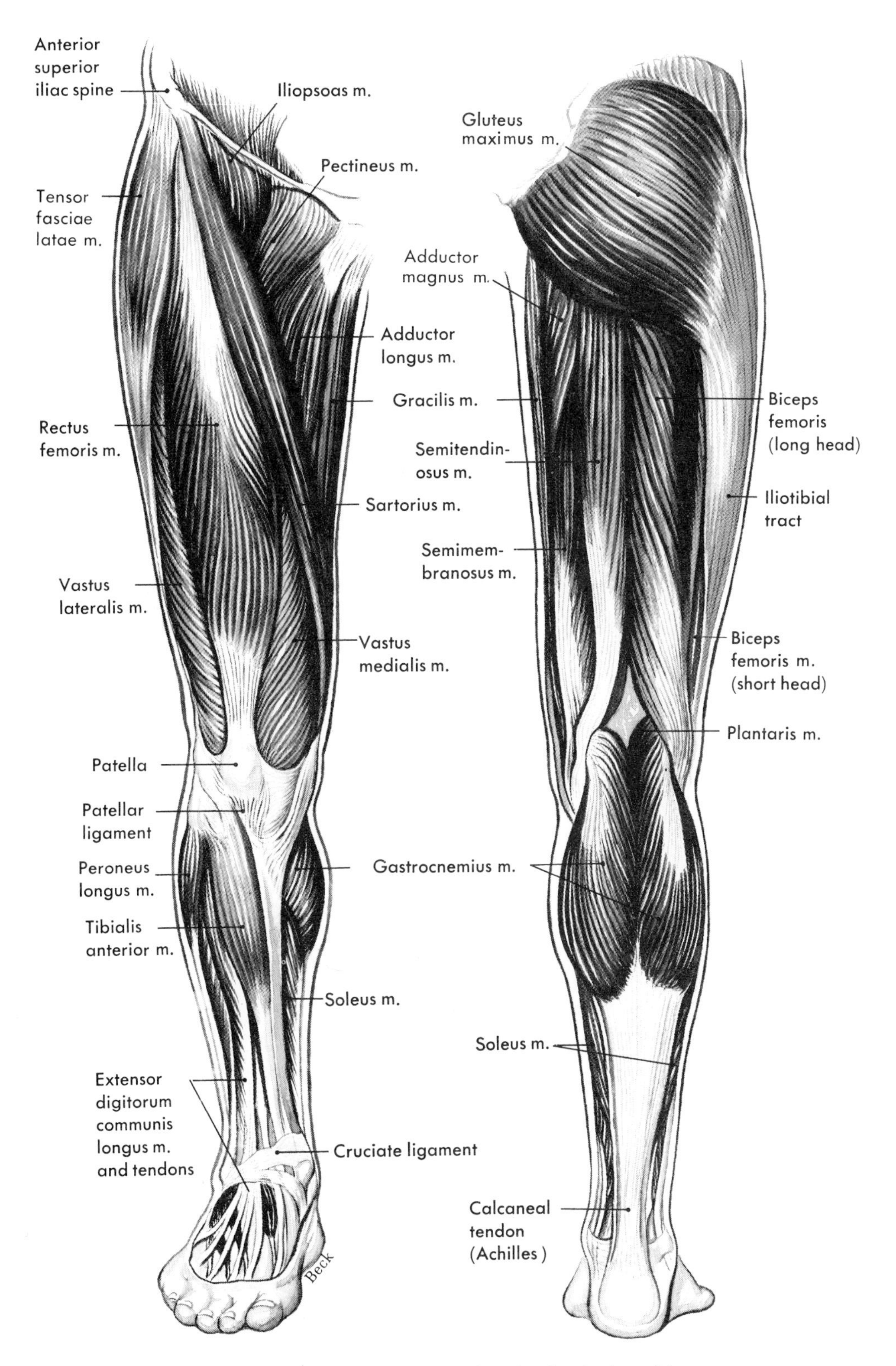

Fig. 4-7 Superficial muscles of right thigh and leg.

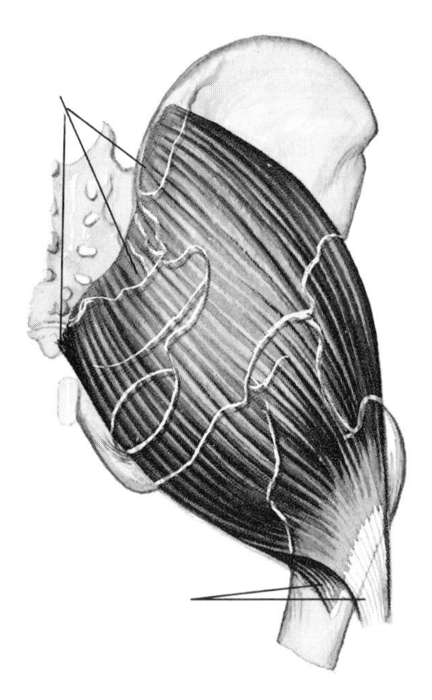

Fig. 4-8 Gluteus maximus muscle.

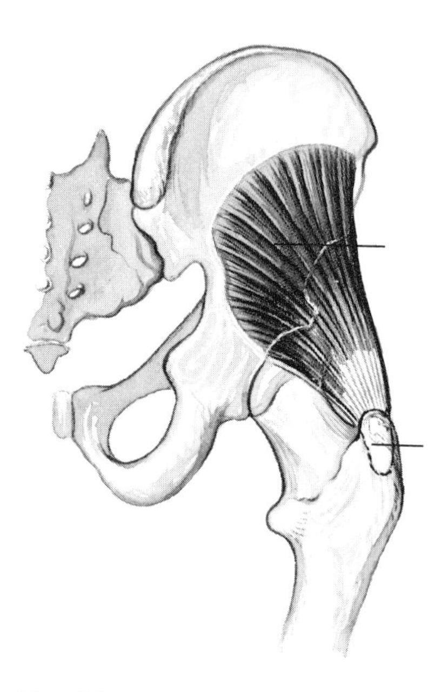

Fig. 4-9 Gluteus minimus muscle.

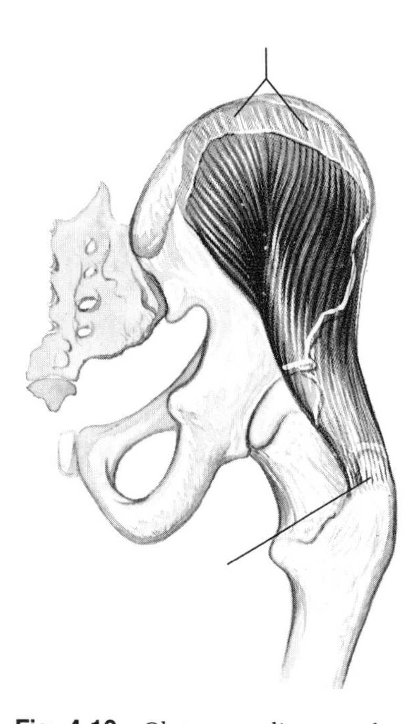

Fig. 4-10 Gluteus medius muscle.

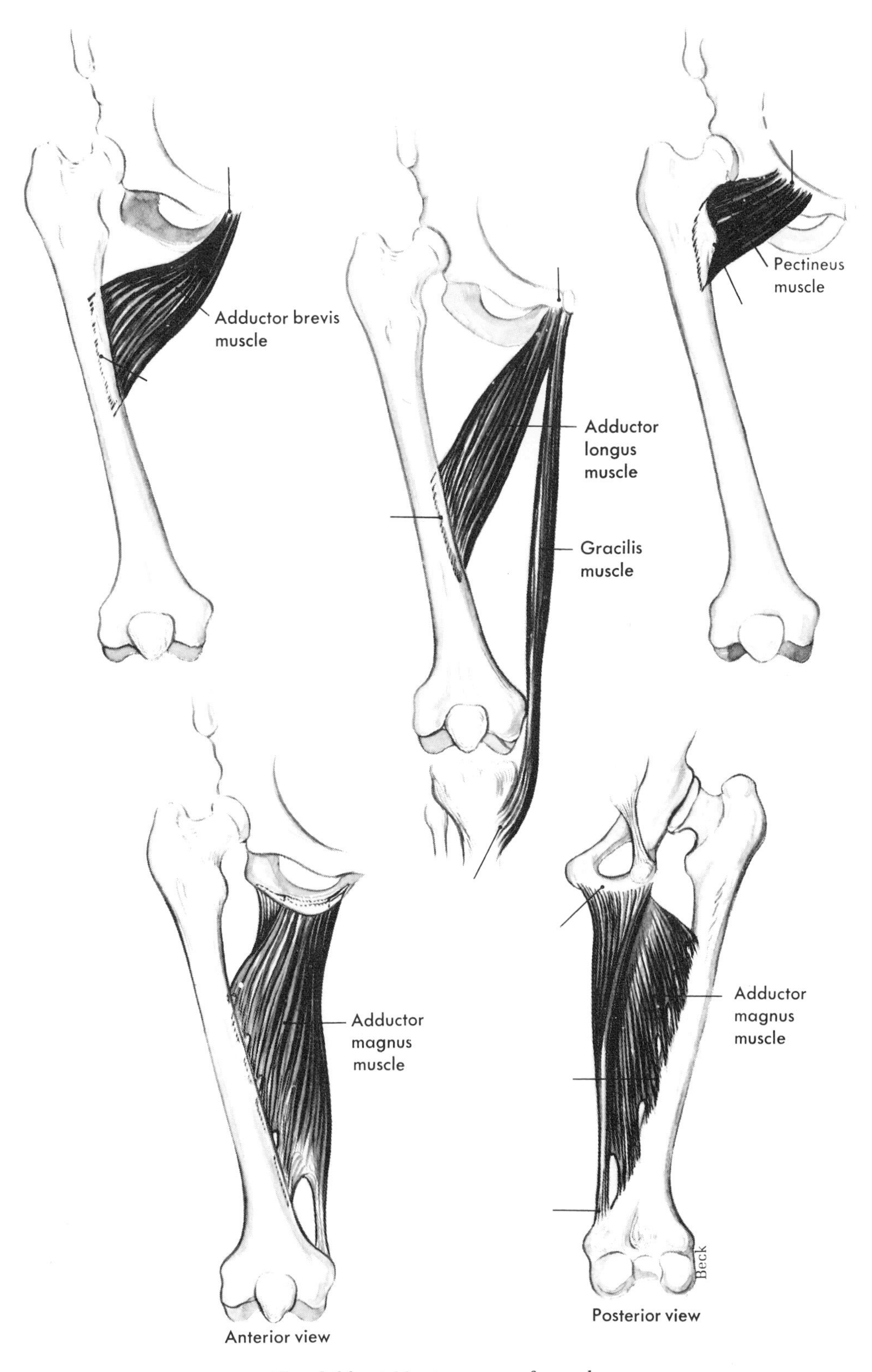

Fig. 4-11 Adductor group of muscles.

Structure, location, and actions of skeletal muscles—cont'd

Procedure E—Examination of muscles that move lower leg

Equipment

Skeleton

Problems

1 What bone or bones serve as the insertion bones for all muscles that act as prime movers of the lower leg? The prime movers we shall investigate are the quadriceps femoris and the hamstring groups of muscles.
2 At which joint does movement occur whenever the lower leg moves?
3 Where is each of these groups of muscles located with relation to bones and joints?
4 What movement of the lower leg does contraction of each of these groups of muscles produce?

Collection of data

1 Study Fig. 4-12 showing the quadriceps femoris group of muscles—rectus femoris, vastus lateralis, vastus medialis, and vastus intermedius—to try to determine:
   a What bone lies under the body of the muscles
   b On what bone or bones they probably have their origins
   c On what bone or bones they probably have their insertions
2 Locate the *rectus femoris* approximately on the skeleton and on your own body.
3 Deduce the main movement of the lower leg produced by contraction of the quadriceps femoris group of muscles. Answer questions 1 to 6 under conclusions.
4 Repeat steps 1, 2, and 3 for the *hamstring* group of muscles—biceps femoris, semitendinosus, and semimembranosus (Fig. 4-13). Answer questions 7 to 11 under conclusions.

Conclusions

1 Identify the origin and insertion bones for the quadriceps femoris and hamstring groups of muscles shown in Figs. 4-12 and 4-13 by writing the word origin or insertion on the appropriate label lines.
2 What bone(s) necessarily serve as the insertion bone(s) for muscles that act as prime movers of the lower leg?

____

3 At which joint does movement occur whenever the lower leg moves?

____

4 What type joint is this?

____

5 What group of muscles forms the flesh on the anterior surface of the thigh?

____

6 What movement of the lower leg results from contraction of the rectus femoris muscle?

____

7 Which, if any, of the muscles that move the lower leg extend across two joints? Which joints are they?

____

____

____

8 Imagine the knee joint held rigid and the rectus femoris muscle contracting. What movement of what part of the body would result?

____

9 What group of muscles forms the flesh of the posterior surface of the thigh?

____

10 What movement of the lower leg results

from contraction of the hamstring group of muscles?

____

11 Imagine the knee joint held rigid and the hamstrings contracting. What movement of what part of the body would result?

____

12 What group of muscles serves both as extensors of the lower leg and as flexors of the thigh?

____

13 What group of muscles serves both as flexors of the lower leg and as extensors of the thigh?

____

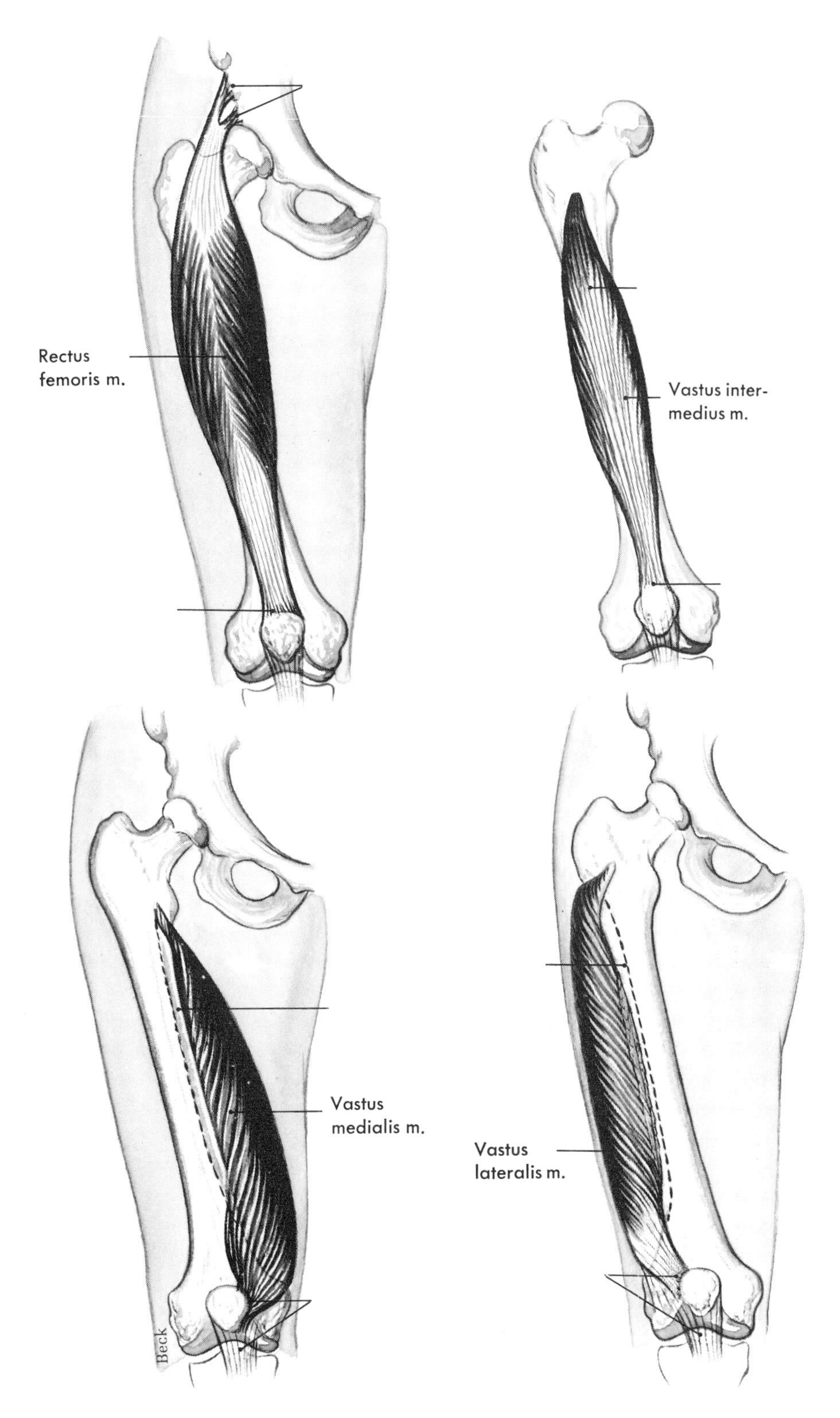

Fig. 4-12 Quadriceps femoris group of muscles.

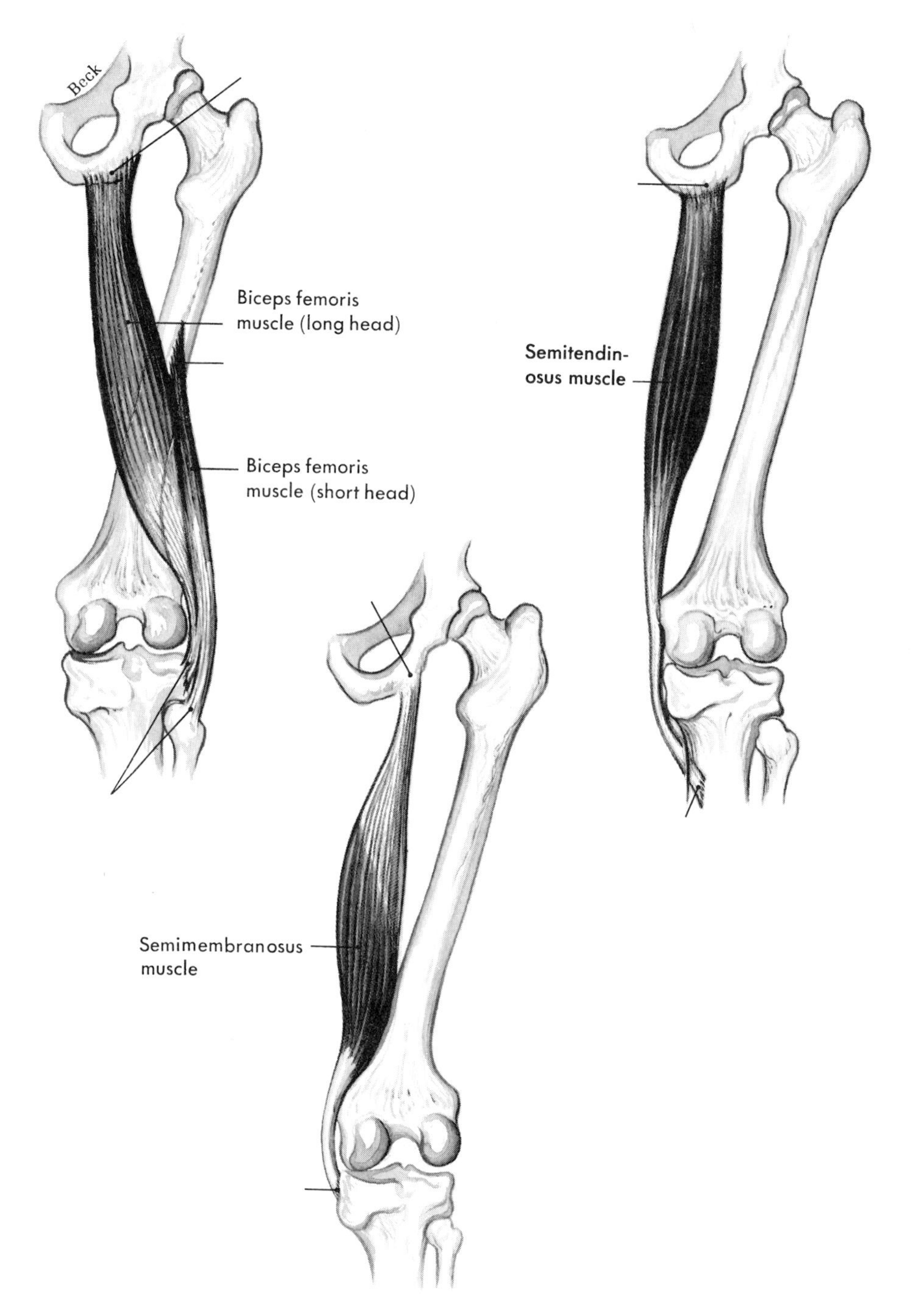

Fig. 4-13 Hamstring group of muscles.

■ Structure, location, and actions of skeletal muscles—cont'd

Procedure F—Examination of muscles that move head, trunk, and abdominal wall

Equipment

Skeleton

Problems

1 What movement of the head does the trapezius produce?
2 What movement of the head does the sternocleidomastoid produce?
3 What movement of the trunk does the rectus abdominis produce?
4 What movement of the trunk does the sacrospinalis produce?
5 Do abdominal muscles contract during breathing? If so, when—during inhalation or exhalation?

Collection of data

1 Examine the trapezius muscle as shown in Figs. 4-1 and 4-2. Note which bones it seems attached to. Answer question 1 under conclusions.
2 Examine Fig. 4-14. Note which bones the sternocleidomastoid muscle attaches to. Answer question 2 under conclusions.
3 Examine Fig. 4-15. Note which bones the rectus abdominis attaches to. Answer question 3 under conclusions.
4 The sacrospinalis muscle is a large complex muscle with extensions all the way up the back from the pelvis to the skull. Some sacrospinalis fibers have their origin on the sacrum, the iliac crest, the lumbar vertebrae, and the lower two thoracic vertebrae and have their insertions on the lower six or seven ribs. Answer questions 4 to 8 under conclusions.
5 Stand erect with one hand resting on your abdomen while you perform the following actions. Note any movement of your abdominal wall during each action.
   a Breathe in and out normally several times.
   b Breathe in normally, then breathe out as much air as you possibly can.
   c Breathe in as much air as you possibly can, then breathe out only a normal amount.

Conclusions

1 When the trapezius muscle contracts and the occipital bone serves as its insertion, the movement it produces is flexion? extension? of the head.

____

2 When the two sternocleidomastoid muscles contract with the temporal bones acting as their insertions, the movement they produce is flexion? extension? of the head.

____

3 When both rectus abdominis muscles contract with the ribs and sternum acting as their insertion, they produce flexion? extension? of the trunk.

____

4 Contraction of the sacrospinalis muscle produces flexion? extension? of the trunk.

____

5 Bowing the head, as in prayer, is produced by contraction of the flexor muscles of the head—viz., the . . . .

____

6 Bending forward from the waist is produced by contraction of the flexor muscles of the trunk, one of which is the . . . .

____

7 One function of the . . . . muscle is to serve as an antagonist to the sterno-

cleidomastoid muscles in head movements.

---

8 Because it holds the spine erect (extended? flexed?), the .... is also called the erector spinae muscle.

---

9 Does contraction of the abdominal muscles move the abdominal wall in or out?

---

10 From your observations in step 5 under collection of data, do you conclude that abdominal muscles contract most strongly during forced inspiration or forced expiration?

---

11 If you wanted to improve your posture by "flattening" your abdomen, which of the following simple exercises should you practice daily—maximal inspiration followed by normal expiration or normal inspiration followed by maximal expiration?

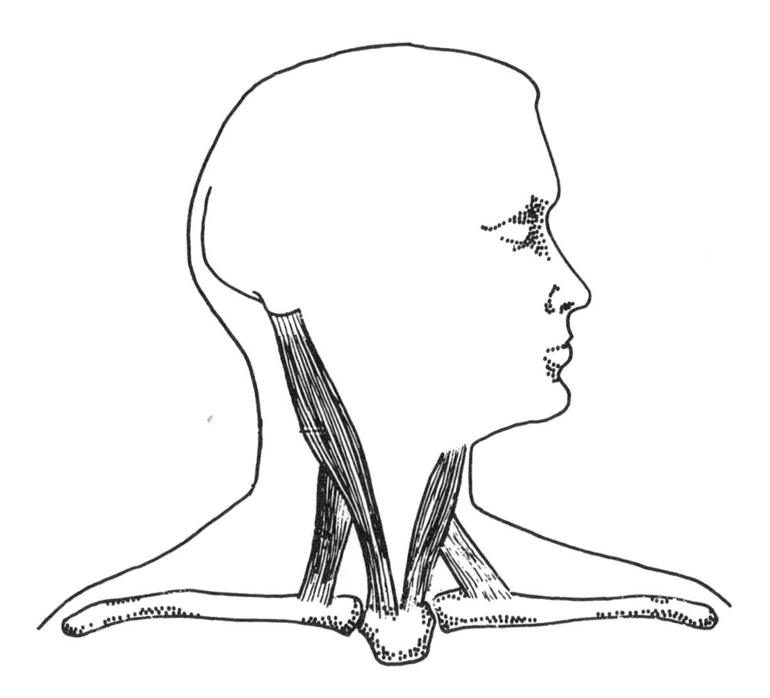

Fig. 4-14 Sternocleidomastoid.

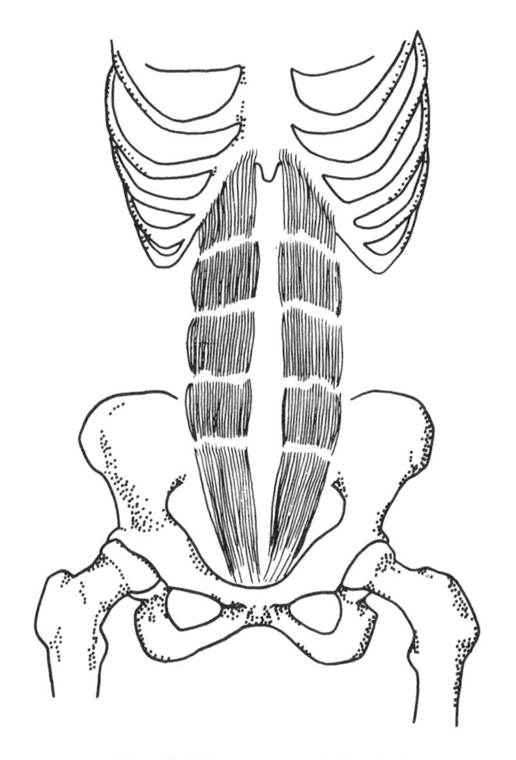

Fig. 4-15 Rectus abdominis.

Muscle physiology

Procedure A—Experiment demonstrating isometric and isotonic contractions

Equipment

No special equipment needed

Problems

1 What is an isometric contraction?
2 How does an isometric contraction differ from an isotonic contraction?

Collection of data

1 Place the palm of your left hand on the undersurface of a desk or table top. Push up on it while you have your right hand cupped over the anterior surface of your left upper arm so that you can feel the muscle there undergo an isometric contraction. Answer questions 1 to 4 under conclusions.
2 Demonstrate an isotonic contraction in this way—start with your left forearm resting on a table. Watch the anterior surface of your left upper arm while you slowly bend your elbow and move your left forearm up toward the upper arm. An isotonic contraction of the muscle on the anterior surface of your upper arm produces this movement. Answer questions 5 and 6 under conclusions.

Conclusions

1 The biceps brachii? triceps brachii? is located on the anterior surface of the upper arm.

____

2 What change did you notice in the firmness of this muscle when you pushed up against the undersurface of the table top?

____

3 Did your hand or forearm move as you pushed up against the table?

____

4 In view of your answer to question 3, do you deduce that this muscle's fibers shortened as you pushed up against the table top?

____

5 If a muscle contraction produces movement, you can be sure that this is an isometric? isotonic? contraction.

____

6 If a muscle's tone (firmness or tension) increases during contraction but its length does not change, you can be sure that this is an isometric? isotonic? contraction.

____

7 Which kind of exercises do you deduce would be more effective for developing muscle strength—isotonic or isometric?

____

8 Which kind of exercises do you deduce would be more effective for promoting flexibility and preventing joint and muscle stiffness—isotonic or isometric?

____

■ Muscle physiology—cont'd
Procedure B—Films

"Human body, the muscular system," color, 13½ min; Modern Film Rentals, 2323 New Hyde Park Road, New Hyde Park, N. Y. 11040 (rental, $8). Shows three types of muscle, explains structure and function of voluntary muscle. Laboratory demonstration illustrates role of nerve impulse in muscle contraction. Also explains role of ATP in movement.

"Muscle contraction and oxygen debt," color, 16 mm, 16 min; McGraw-Hill Films, Dept. WP, 330 West 42nd St., New York, N. Y. 10036 (purchase, $215; rental, $12.50).

"Microelectrodes in muscle," color, 19 min; John Wiley & Sons, Dept. SN, 605 3rd Ave., New York, N. Y. 10016 (purchase, $225; rental, $30). Demonstrates intracellular microelectrode technique to measure electrical potential across membrane of frog muscle cell.

■ Self-test

Consult the diagrams in Fig. 4-16 and answer the following questions.

1 Diagram **G**? **H**? **I**? illustrates flexion of the thigh.

____

2 The prime flexor of the thigh is the .... muscle.

____

3 Diagram **I**? **J**? illustrates flexion of the lower leg.

____

4 A muscle that flexes the lower leg has to be located on the anterior? posterior? surface of the lower extremity.

____

5 A muscle that flexes the lower leg must necessarily have its origin above? below? on? the lower leg.

____

6 The prime flexor of the lower leg is the .... group of muscles.

____

7 Diagram **K**? **L**? illustrates flexion of the head.

____

8 A muscle that flexes the head has to be located on the anterior? posterior? surface of the body.

____

9 The prime flexor of the head is the .... muscle.

____

10 Diagram **O**? **P**? illustrates flexion of the trunk.

____

11 A muscle that flexes the trunk has to be located on the anterior? posterior? surface of the body.

____

12 One of the main flexors of the trunk is the .... muscle.

____

13 Diagram **J** shows flexion? extension? of the lower leg.

____

14 A muscle that produces the movement shown in diagram **J** would have to lie on the anterior? posterior? surface of the lower leg? thigh?

____

____

15 The movement illustrated by diagram **J** is produced by contraction of the .... muscle.

____

16 Diagram **H** shows abduction? extension? flexion? of the thigh.

____

17 Contraction of the gluteus maximus produces the movement illustrated by diagram ....

---

18 Contraction of the gluteus medius and minimus produces the movement illustrated by diagram ....

---

19 Contraction of the latissimus dorsi produces abduction? extension? flexion? of the upper arm.

---

20 The antagonist to the latissimus dorsi in upper arm movements is the .... muscle.

---

21 Extension of the lower arm results from contraction of the .... muscle.

---

22 Based on your study of muscles, which of the following do you conclude? Most flexor muscles lie on the anterior? posterior? surfaces of the body, whereas most extensors lie on the anterior? posterior? surfaces.

---

23 An exception to the principle stated in question 22 is the location of the lower leg extensor? flexor? muscles that lie on the posterior surface of the body.

---

24 The location of the extensor muscles of the lower leg? thigh? is not an exception to the principle stated in question 22.

---

25 The location of the extensor muscles of the lower leg is? is not? an exception to the principle stated in question 22.

Fig. 4-16

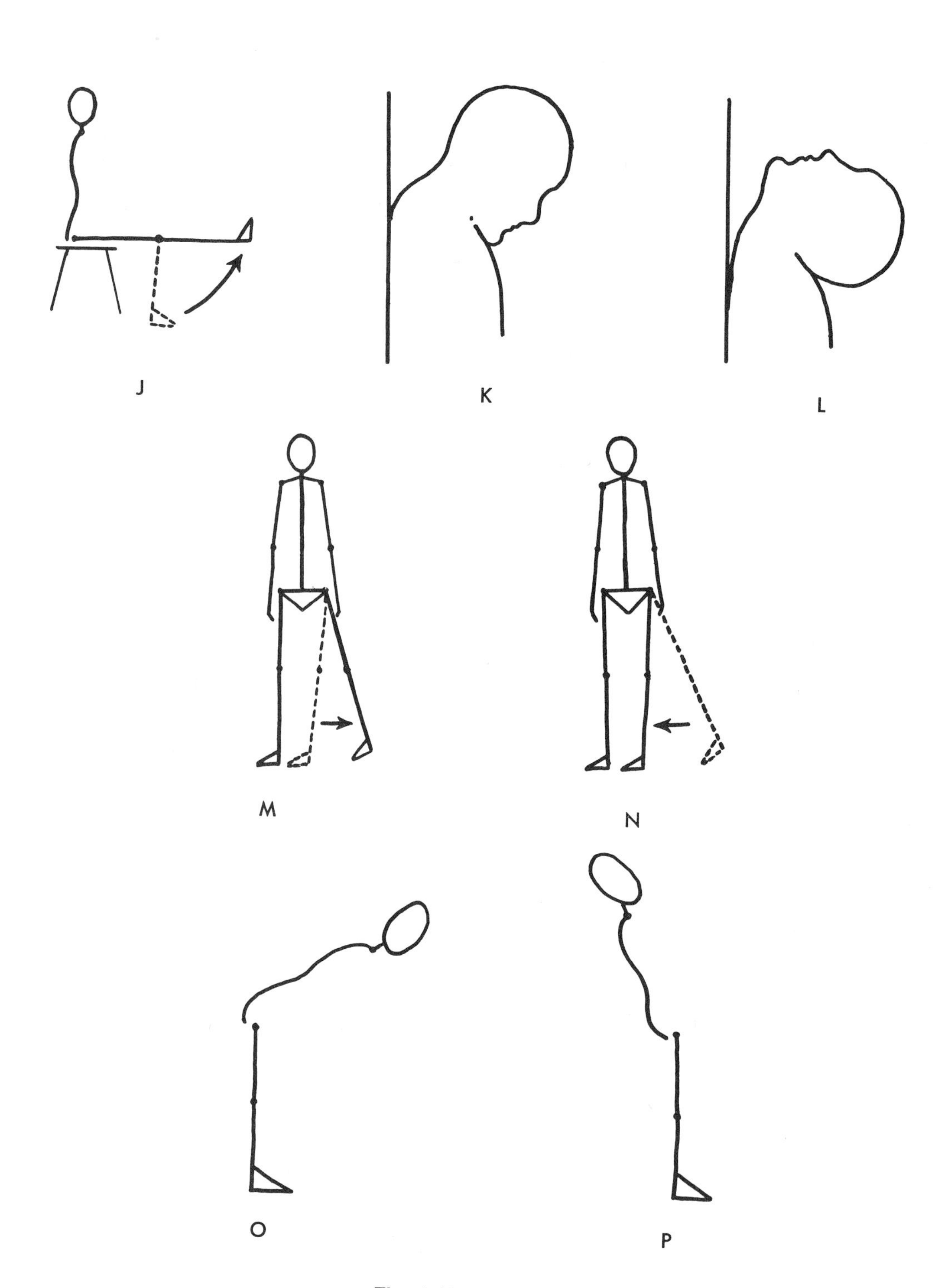

Fig. 4-16, cont'd

UNIT THREE

HOW THE BODY CONTROLS AND INTEGRATES ITS FUNCTIONS

5 The somatic nervous system

■ Introduction

Procedure—Films

"Nervous control of behavior," color, 16 mm, 16 min; McGraw-Hill Films, Dept. WP, 330 West 42nd St., New York, N. Y. 10036 (purchase, $215; rental, $12.50). Examines in detail structure of nerve cells, vertebrate brain, and spinal nerves. Demonstrates Helmholtz's experiment to measure nerve impulse speed. Also demonstrates spinal reflex in frog.

"The human body: nervous system," color, 13½ min; Modern Film Rentals, 2323 New Hyde Park Road, New Hyde Park, N. Y. 11040 (rental, $8). Shows various neurons and main organs of nervous system; emphasizes basic functions.

■ Spinal cord and spinal nerves

Procedure A—Examination of spinal cord and spinal nerves

Equipment

1 Fresh sheep cords soaked in 5% formalin overnight
2 Preserved human cord
3 Model of cross section of human cord, spinal nerve roots attached
4 Model of spinal nerve peripheral branches
5 Charts and illustrations

Problems

1 What gross structural features characterize the spinal cord and spinal nerves?
2 In what gross structure are dendrites of sensory neurons located?
3 In what gross structure are cell bodies of sensory neurons located?
4 In what gross structures are axons of sensory neurons located?
5 In what gross structures are dendrites of somatic motor neurons located?
6 In what gross structures are cell bodies of somatic motor neurons located?
7 In what gross structures are axons of somatic motor neurons located?

Collection of data

1 Examine illustrations, specimens, and model of the spinal cord to identify the following:
   a Anterior and posterior columns of gray matter
   b Anterior, lateral, and posterior columns of white matter
   c Anterior and posterior roots of spinal nerves
   d Spinal ganglia
2 Study Fig. 5-1 to collect the data you need to answer the questions under conclusions.

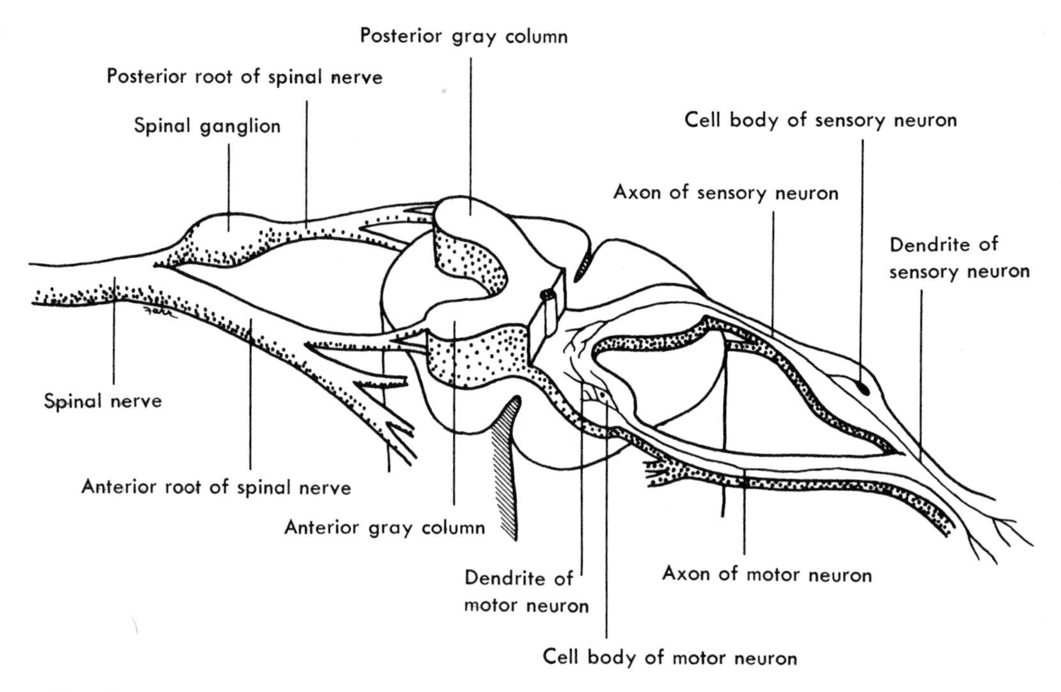

Fig. 5-1 Cross section of spinal cord with its attached spinal nerve roots and spinal nerves.

Conclusions

1 The H-shaped inner core of the spinal cord consists of gray? white? matter.

____

2 As Fig. 5-1 shows, spinal nerves attach directly? indirectly? to the spinal cord.

____

3 Attached to each of the thirty-one segments of the spinal cord is one pair of spinal nerves. Each spinal nerve attaches to the cord by means of an anterior and a posterior ....

____

4 One can easily identify the anterior? posterior? root of a spinal nerve by a small swelling on it called the spinal ....

____

____

5 Dendrites of sensory neurons are located in the posterior roots of spinal nerves? in spinal nerves?

____

6 Cell bodies of sensory neurons are located in the anterior gray columns of the cord? in the spinal ganglia?

____

7 Axons of sensory neurons are located in the anterior? posterior? roots of spinal nerves.

____

8 Neurons whose axons supply skeletal muscles are called somatic motoneurons. Dendrites of somatic motoneurons are located in the anterior roots of spinal nerves? in the anterior gray columns of the cord?

____

9 Cell bodies of somatic motoneurons are

located in the anterior gray columns of the cord? in spinal ganglia?

____

10 Axons of somatic motoneurons are located in the anterior? posterior? roots of the spinal nerves and in the spinal nerves themselves.

____

11 Spinal nerves contain dendrites of sensory neurons? of somatic motoneurons?

____

12 Spinal nerves also contain axons of sensory neurons? of somatic motoneurons?

____

13 Spinal ganglia are composed of the cell bodies of sensory neurons? of somatic motoneurons?

____

14 The only cell bodies of neurons located in the anterior gray columns of the cord are those of sensory neurons? of somatic motoneurons?

____

15 Posterior roots of spinal nerves contain axons of sensory neurons? of somatic motoneurons?

____

16 Anterior roots of spinal nerves contain axons of sensory neurons? of somatic motoneurons?

____

17 Spinal nerves conduct impulses both toward and away from the spinal cord by means of the neuron processes they contain—toward the cord by means of axons? dendrites? of sensory neurons and away from the cord by means of axons? dendrites? of motoneurons. In other words, spinal nerves are "mixed nerves"—i.e., they are neither purely sensory nor purely motor.

____

____

18 The only way nerve impulses can reach skeletal muscles is via somatic motoneurons. The polio virus makes these neurons unable to conduct impulses because it attacks their cell bodies, which are located in the ....
Result: Paralysis of cardiac muscle? smooth muscle? skeletal muscle?

____

____

19 Based on the data you have collected in this procedure, you would reason logically that disease of the posterior roots of spinal nerves could? could not? produce a loss of sensation.

____

■ Spinal cord and spinal nerves—cont'd

Procedure B—Spinal cord functions; stretch reflexes

Equipment

1 Rubber hammer
2 Textbook of anatomy and physiology

Problems

1 What is a receptor?
2 What is an effector?
3 What is a synapse?
4 What is a reflex arc?
5 What is a reflex?
6 What is a stretch reflex?
7 What kind of reflex arc mediates a stretch reflex?
8 What constitutes the reflex center of a two-neuron reflex arc?
9 Where are the reflex centers of two-neuron reflex arcs located?

Collection of data

1 Ask your laboratory partner to sit on a chair or stool, cross one leg over the other, and close her eyes. Using a moderate amount of force, tap the skin over her patellar tendon with a rubber hammer or with the edge of your hand. Note the movement of her lower leg in response to this stimulus.
2 Support your partner's elbow in your left hand, and with the rubber hammer tap the tendon of her triceps muscle sharply just above the elbow. Note the movement of her forearm in response to this stimulus.
3 Ask your partner to kneel on a chair with her ankles and feet projecting over the edge. Sharply tap her Achilles tendon with a rubber hammer. Note the movement of her foot in response to this stimulus.
4 Nerve impulse conduction is initiated by an adequate stimulus acting on receptors (distal ends of sensory neurons).
5 Dendrites conduct impulses to the cell body of a neuron, and axons conduct impulses away from the cell body.
6 Look up the words synapse and effector in your textbook.
7 According to one definition, a reflex arc is any neural pathway that conducts impulses from receptors to effectors.
8 Study the right side of Fig. 5-1. It shows the neurons that make up the simplest kind of reflex arc.
9 The center of the reflex are shown in Fig. 5-1 is the synapse in it.

Conclusions

1 Impulse conduction usually starts in the distal ends of axons? dendrites? of motor? sensory? neurons.

____

____

2 The distal ends of sensory neurons are called ....

____

3 In any neuron, its axon? dendrite? conducts impulses to its cell body, and its .... conducts impulses away from the cell body.

____

____

4 A neural pathway that conducts impulses from a receptor to an effector is called a ....Conduction over such a pathway causes the effector to respond; this response is called a reflex.

____

5 Whereas a receptor is part of a motor? sensory? neuron, an effector is either a .... or a ....

____

____

____

6 The simplest kind of reflex arc consists, as Fig. 5-1 shows, of ....?

7 Impulse conduction over any reflex arc is initiated by stimulation of .... and results in a response by .... The response is called a ....

8 Tendons are fibrous cords that insert muscles into bones. As you can deduce from its location, the patellar tendon inserts the rectus femoris muscle into the .... Tapping a muscle's tendon stretches both the tendon and its muscle.

9 Stretching a muscle stimulates receptors in it called muscle spindles. This initiates conduction over a two-neuron? three-neuron? reflex arc of the type shown in Fig. 5-1. Since this particular arc's motoneuron axon terminates in the muscle that was stretched, conduction by the arc stimulates that muscle to respond by contracting. Its contraction produces a movement that is referred to as a stretch ....

10 The extension? flexion? of the lower leg that you observed in step 1 under collection of data is an example of a stretch reflex.

11 The effector that responded to produce the stretch reflex observed in step 1 under collection of data was the ....

12 What stretch reflex did you elicit by carrying out step 2 under collection of data?

13 What effector produced the stretch reflex observed in step 2 under collection of data?

14 What stretch reflex did you elicit by carrying out step 3 under collection of data?

15 What effectors produced the stretch reflex observed in step 3 under collection of data?

16 All stretch reflexes are mediated by conduction over the simplest neural pathways—viz., over ....

17 As Fig. 5-1 shows, the reflex center for a stretch reflex arc lies in the gray? white? matter of the cord.

18 The reflex center for a stretch reflex arc or for any two-neuron arc consists of the places of contact between the axons of sensory neurons and the dendrites or cell bodies of motoneurons. Or, in one word, the reflex center for a stretch reflex consists of ....

■ Spinal cord and spinal nerves—cont'd
Procedure C—Spinal cord functions; withdrawal reflexes

Equipment

Sterilized needle

Problems

1. What type reflex arc mediates withdrawal reflexes?
2. What microscopic structures constitute the reflex centers of these reflex arcs?
3. Where are these reflex centers located?

Collection of data

1. Ask your laboratory partner to sit down with her eyes closed and to place one hand palm up on the desk. Using a sterilized needle, suddenly prick the index finger of this hand. Note the resulting movement of the forearm, a response appropriately called a withdrawal reflex.
2. Study Fig. 5-2. It diagrams the type reflex arc that mediates withdrawal reflexes.

Conclusions

1. The withdrawal reflex you elicited in this experiment consisted of extension? flexion? of the forearm.

____

2. This experiment indicates that a withdrawal reflex may be elicited by stimuli that give rise to a sensation of . . . .

____

3. The effectors for this or for any withdrawal reflex are branching? smooth? striated? muscle cells.

____

4. The biceps brachii? triceps brachii? muscle was an effector for the withdrawal

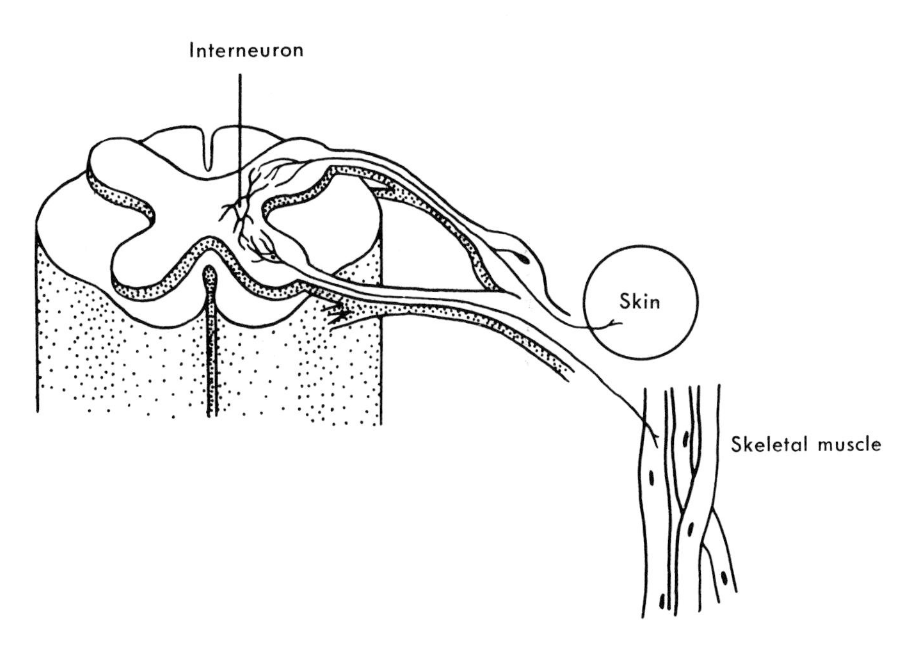

Fig. 5-2 Three-neuron reflex arc that mediates withdrawal reflexes.

reflex you elicited by pricking your laboratory partner's finger.

____

5 Two-neuron? three-neuron? arcs mediate stretch reflexes. As Fig. 5-2 shows, ... arcs mediate withdrawal reflexes.

____

____

6 As Fig. 5-2 shows, the reflex centers for three-neuron arcs consist of .... In contrast, the reflex centers for two-neuron arcs consist of synapses between the axons of motoneurons? of sensory neurons? and the dendrites or cell bodies of ....

____

____

____

7 The stretch reflexes you observed in Procedure B—and, in fact, most stretch reflexes—are produced by contraction of extensor? flexor? muscles.

____

8 Most withdrawal reflexes, like the one you observed in this procedure, are produced by contraction of extensor? flexor? muscles.

____

Brain and cranial nerves

Procedure A—Examination of coverings of brain

Equipment

1 Fresh sheep head from slaughterhouse
2 Preserved human brain with meninges intact
3 Charts and illustrations

Problems

1 What features characterize the coverings around the brain?
2 What purpose do these coverings serve?

Collection of data

1 A triple-layered membrane called the meninges forms a covering that encloses both the brain and the spinal cord. Examine Fig. 5-3 to identify the three layers of meninges. Fill in the blanks for items 1 to 3 under conclusions.
2 Have the sheep head sawed horizontally a short distance above the eyes. Care should be taken to cut just through the bone but not through the underlying membranes.
3 a Remove the "skull cap." Note the membranous lining adherent to the inner surface of the skull bones.
  b Note whether the membrane forming the outer layer of the meninges is strong and tough or delicate and easily torn.
  c Note the channel formed by a double layer of this membrane along the midline of the skull. This is one of the principal veins of the skull. Answer questions 4 and 5 under conclusions.
4 Observe the "cobwebby" middle layer of the meninges on the specimen and in Fig. 5-3.
5 Try to pull away some of the delicate membrane adherent to the outer surface of the brain. Answer questions 6 and 7 under conclusions.

Conclusions

1 The dense (white) fibrous tissue layer of the meninges adheres to the inner surface of the cranial bones; its name is ....

____

2 The .... is the delicate layer of the meninges adherent to the brain.

____

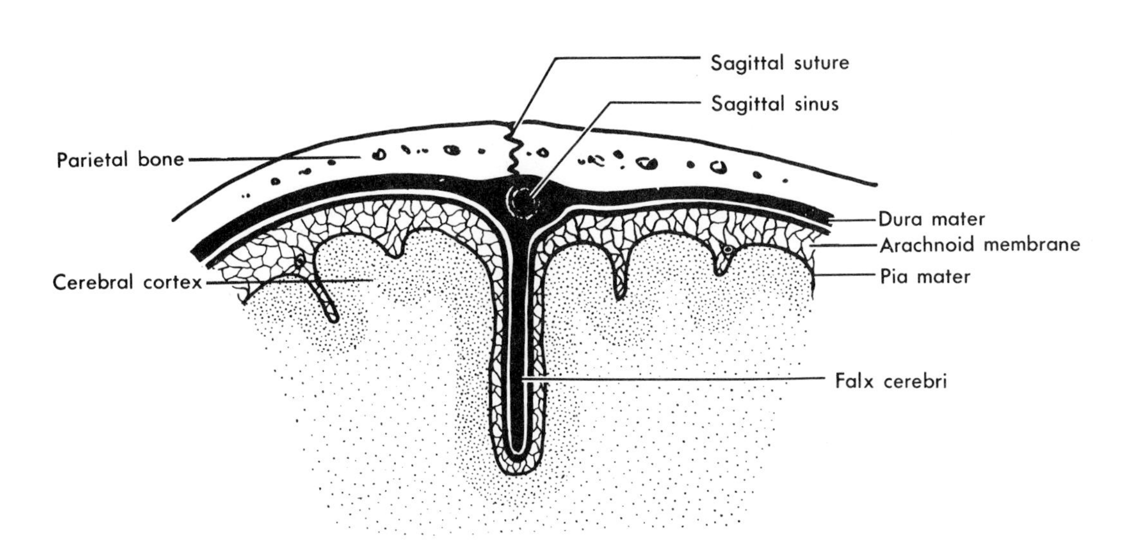

Fig. 5-3 Coronal section through parietal bones, meninges, and cerebrum.

3 The name of the delicate "cobwebby" middle layer of the meninges is the . . . .

---

4 Beneath the sagittal suture lies a vein called, as Fig. 5-3 shows, the sagittal . . . .

---

5 A double layer of the . . . . mater forms the sagittal (or longitudinal) sinus.

---

6 Because the inner? middle? outer? layer of the meninges reminded early anatomists of a cobweb, they named it the . . . . from the Greek word meaning spider.

---

7 The pia mater is the inner? outer? layer of the meninges, the dura mater is the . . . . layer, and the arachnoid lies between them.

■ Brain and cranial nerves—cont'd

Procedure B—Examination of human brain

Equipment

1 Charts and illustrations
2 Dissectible model of human brain
3 Preserved human brains (intact; also sagittal and coronal sections)

Problem

What major gross structures compose the human brain?

Collection of data

1 Examine various illustrations and consult your textbook to:
 a Identify the following parts of the human brain—cerebrum, diencephalon (composed mainly of thalamus and hypothalamus), midbrain, pons, medulla, and cerebellum. Locate each of these structures in either Fig. 5-4 or Fig. 5-5 or in both.
 b Identify these parts of the cerebrum:
  1 Lobes
  2 Convolutions or gyri
  3 Sulci
  4 Cortex
  5 Corpus callosum (locate in Fig. 5-5)
  6 Basal ganglia (or cerebral nuclei; names of main ones are caudate, putamen, and pallidum)
 c Identify two parts of the diencephalon—viz., the thalamus and the mamillary bodies (posterior part of the hypothalamus).
 d Identify the ventricles located in the interior of the brain—viz., the right and left lateral ventricles (not shown in figures), the third ventricle, and the fourth ventricle. Also, identify the cerebral aqueduct.
 e Identify the pituitary gland (hypophysis) and the stalk that attaches it to the undersurface of the brain.
 f Identify the pineal body (epiphysis).
 g Identify some of the cranial nerve attachments to the undersurface of the brain—e.g., the optic nerves and optic chiasma and the olfactory bulbs (tracts). Locate as many as you can in Figs. 5-4 and 5-5.
2 Locate as many as possible of the structures on the dissectible model of the human brain and then on the preserved specimen.

Conclusions

1 Follow the instructions accompanying Fig. 5-4.
2 Follow the instructions accompanying Fig. 5-5.

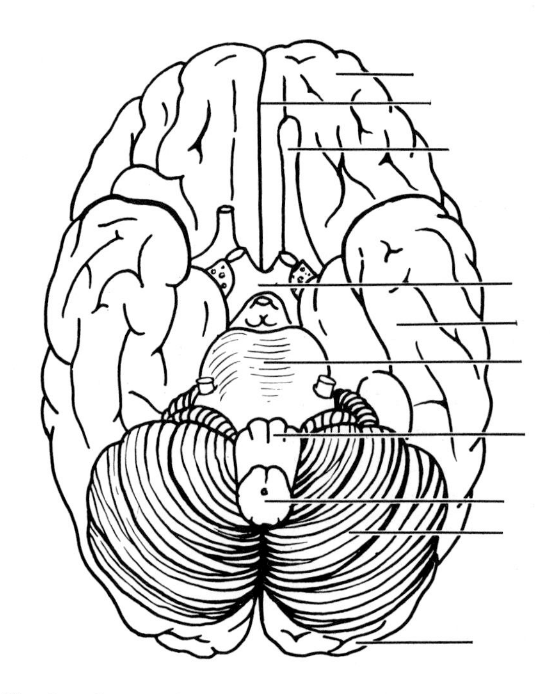

Fig. 5-4 Base of brain. Place each of the following terms on the appropriate label line:

Cerebellum
Frontal lobe of cerebrum
Sagittal (longitudinal) fissure
Medulla
Occipital lobe
Olfactory bulb
Optic chiasma
Pons
Spinal cord
Temporal lobe

Label the following:

Mamillary bodies
Optic nerve
Olfactory nerve

Color the following:

Cerebrum .... red
Cerebellum .... blue
Spinal cord .... } yellow
Medulla oblongata .... } yellow
Pons .... } yellow
Cranial nerves .... } yellow

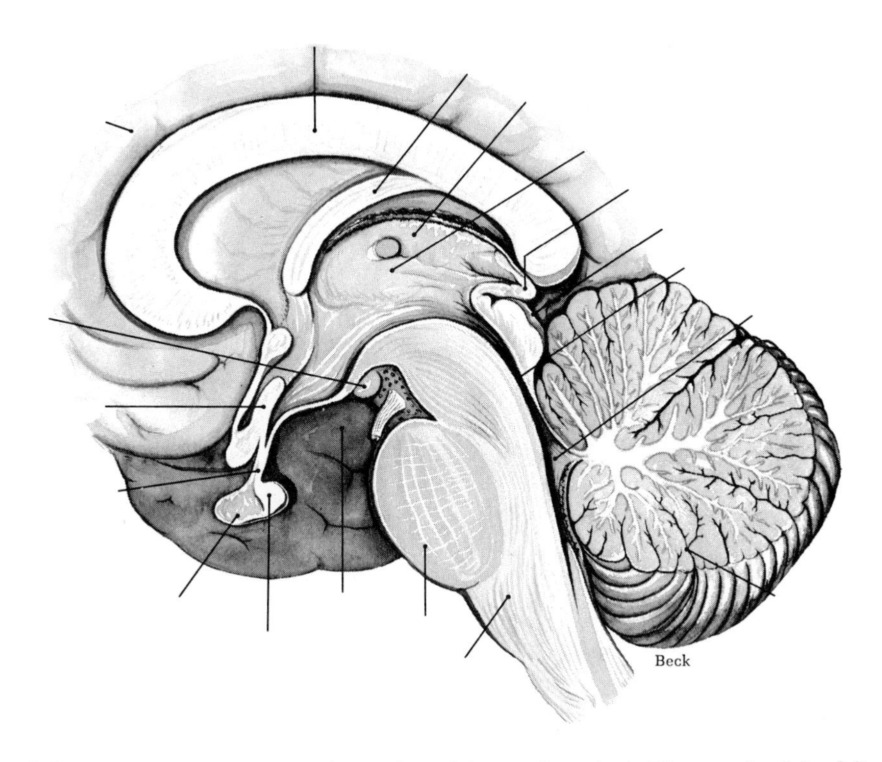

Fig. 5-5 Brain and upper part of spinal cord (sagittal section). Place each of the following terms on the appropriate label line, leaving any other label lines blank:

Anterior lobe of pituitary gland
Cerebellum
Cerebral aqueduct
Colliculi (posterior part of midbrain)
Corpus callosum
Fourth ventricle
Cerebrum
Mamillary body
Medulla oblongata
Optic chiasma
Pineal body
Posterior lobe of pituitary gland
Stalk of pituitary gland
Pons
Thalamus
Third ventricle

Color the following:

Cerebrum .... red
Cerebellum .... blue
Cerebral aqueduct .... } purple
Fourth ventricle ...... } purple
Third ventricle .... } purple
Pituitary gland stalk ... green
Pineal body .... green
Medulla oblongata ..... } yellow
Pons .... } yellow
Midbrain .... } yellow

■ Brain and cranial nerves—cont'd

Procedure C—Dissection of sheep brain

Equipment

1. Fresh sheep brains, preferably soaked in 5% formalin overnight
2. Dissecting instruments
3. Dissecting pans

Problem

What structural features of a human brain are also characteristic of a sheep brain?

Collection of data

1. Before cutting into the brain, examine it externally to identify the structures shown in Fig. 5-4.
2. Place the brain ventral surface down with all parts straight in line. Separate the cerebral hemispheres with your fingers and draw a sharp scalpel straight down through the fissure to section the brain longitudinally. Try to identify the structures shown in Fig. 5-5. Locate the lateral ventricles. Note their shape and size.
3. Make a sagittal section through the cerebrum about midway between the sagittal fissure and the lateral boundary of the cerebrum. This section should enable you to see the corpus striatum (L. for "striped body"). The corpus striatum consists of the basal ganglia named the caudate nucleus and the lenticular nucleus (made up of the putamen and the pallidum)—with alternating stripes of gray matter and white matter (internal capsule) located between them. Note the difference in color between the gray matter that composes the cerebral cortex and the white matter that makes up much of the interior of the cerebrum.

Conclusions

1. Name several structures you observed in the sheep brain that are also present in the human brain.

   ____

   ____

   ____

2. Which part of the sheep brain would you judge is relatively smaller than the same part in the human brain?

   ____

Procedure D—Films

"The human body: the brain," color, 16 min; Modern Film Rentals, 2323 New Hyde Park Road, New Hyde Park, N. Y. 11040 (rental, $8). Shows basic functions of human brain by means of laboratory demonstrations and animations.

"The brain and behavior"; Field Services, Indiana University Audio-Visual Center, Bloomington, Ind. 47401.

Procedure E—Sensory paths to cerebral cortex from periphery

Equipment

Textbook of anatomy and physiology

Problems

1 What is a tract?
2 Which spinal cord tracts conduct impulses that result in:
   a Pain or temperature sensations?
   b Kinesthesia (sense of the position and movement of body parts)?
   c Discriminating touch and pressure sensations; specifically, precise localization of touch and pressure sensations, sense of vibrations, and stereognosis (i.e., sense of size, shape, and texture of objects)?
   d Crude touch and pressure sensations (sensing merely that something is touching or pressing on you)?
3 Which columns of the spinal cord contain tracts that conduct impulses responsible for:
   a Discriminating touch and pressure sensations?
   b Kinesthesia?
   c Crude touch and pressure sensations?
   d Pain and temperature sensations?
4 What two kinds of sensations result from conduction by fibers of the medial lemniscus?

Collection of data

1 Blindfold your laboratory partner or ask her to close her eyes; then move one of her fingers or some other part of her body. Ask her the following questions:
   a Did I touch you? If so, tell me as exactly as you can, where you were touched.
   b Did I move any part of your body? If so, tell me which part I moved, in which direction I moved it, and in which position I left it?
2 While your partner is blindfolded or has her eyes closed, place a small object—tennis ball, pencil, ballpoint pen, rubber band, paper cup, or whatever—in her hand. Ask her to describe the object's size, shape, and texture as accurately as possible and to identify it by name if she can.
3 Study Fig. 5-6. Use data you collect from this figure to answer questions 1 and 2 under conclusions.
4 Consult your textbook to collect data needed for answering the questions stated under problems. Answer questions 3, 4, and 5 under conclusions.
5 Study Fig. 5-7. Use data you collect from this figure to answer questions 6 and 7 under conclusions.

Conclusions

1 Impulse conduction from the skin or muscles or any part of the body to the spinal cord requires the functioning of at least one? two? three? sensory neuron(s)?

____

2 Spinothalamic tracts, like other spinal cord sensory tracts, consist of axons of sensory neurons I? II? III?

____

3 A bundle of axons in the brain or in the cord is called a ....

____

4 The following nicknames would be appropriate for which spinal cord tracts?

a "Discriminating touch pathways" for

______

b "Crude touch pathways" for

______

c "Hot and cold pathways" for

______

5 What spinal cord tracts must have functioned if, in experiments 1 and 2 under collection of data, your laboratory partner sensed:

a The exact place you touched her?

______

b That you moved a part of her body?

______

c The position in which you left the part of her body after moving it?

______

d The size, shape, and texture of the object you placed in her hand and was able to identify it?

______

6 In which columns of the cord are the following located:

a Fasciculi cuneatus and gracilis?

______

b Spinothalamic tracts?

______

7 Which columns of the cord contain the:

a Pathways for sensing that something is touching you someplace?

______

b Pathways for vibratory sensations?

______

c Pathways for stereognosis?

______

d Pathways for precise localization of touch and pressure sensations?

______

e Pathways for sensing the position of a body part without looking to see where it is?

______

f "Pain pathways"?

______

g "Kinesthetic pathways"?

______

h "Hot and cold pathways"?

______

8 The tract in the brain that conducts discriminating touch and pressure impulses from the medulla to the thalamus is called the ....

______

9 The medial lemniscus fibers relay impulses from the fasciculi cuneatus and gracilis? the spinothalamic tracts? to the thalamus.

______

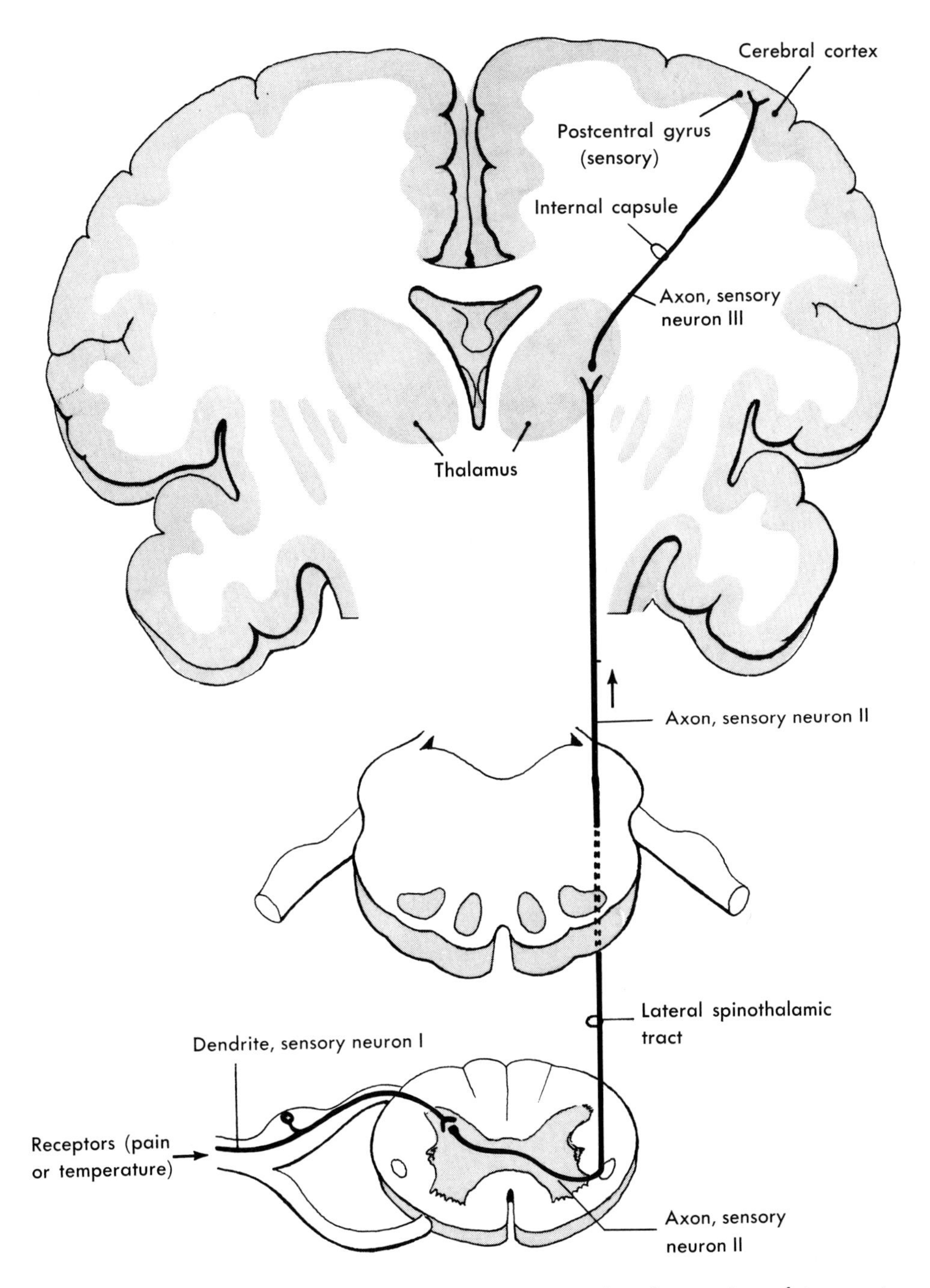

Fig. 5-6 Lateral spinothalamic tract relays sensory impulses from pain and temperature receptors.

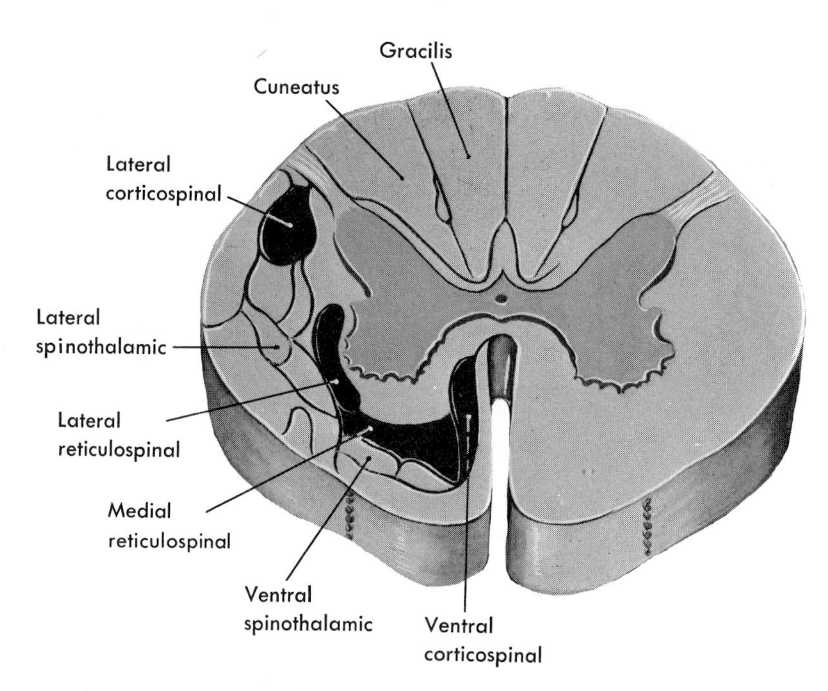

Fig. 5-7 Location in spinal cord of some major projection tracts.

■ Brain and cranial nerves—cont'd

Procedure F—Motor paths from cerebral cortex to skeletal muscles (somatic effectors)

Equipment

Textbook of anatomy and physiology

Problems

1 Pyramidal tract fibers originate from cell bodies of neurons located in what organ?
2 In what organ do pyramidal tract fibers terminate?
3 Where are pyramidal tracts located in the spinal cord?
4 What is an "upper motoneuron"?
5 What is a "lower motoneuron"?
6 What functions do pyramidal tracts serve?

Collection of data

1 Neural paths from the cerebral cortex to the skeletal muscles are varied and complex. Fig. 5-8 shows the simplest and most direct of these.
2 Study Fig. 5-8, and then answer questions 1, 2, 5, and 6 under conclusions.
3 Study Fig. 5-7, and then answer questions 3 and 4 under conclusions.

Conclusions

1 Fig. 5-8 shows axons of the pyramidal tracts originating from neuron cell bodies located in the cord? cortex of the cerebrum?

______

2 Pyramidal tract fibers terminate, as Fig. 5-8 shows, in the cord? cortex of the cerebrum?

______

3 Fig. 5-7 identifies pyramidal tracts by another name, one that indicates the locations of their cells of origin and of their axon terminals. This other name for pyramidal tracts is ....

______

4 As Fig. 5-7 shows, both the anterior (ventral) and the .... white columns of the cord contain corticospinal tracts.

______

5 From Fig. 5-8, you can deduce the reason the tracts shown there are called pyramidal tracts—because the crossing over or .... of their fibers forms the pyramids of the medulla.

______

6 The pyramidal tracts whose fibers decussate in the medulla are appropriately called crossed? direct? pyramidal tracts.

______

7 Pyramidal tract fibers that decussate in the medulla descend in the lateral white columns of the cord. Another name, therefore, for crossed pyramidal tracts is the one used in Fig. 5-7—i.e., ...

______

8 Pyramidal tracts whose fibers do not decussate in the medulla are appropriately called crossed? direct? pyramidal tracts.

______

9 Whereas each lateral? ventral? corticospinal tract consists of axons that have not decussated in the medulla, each ...... corticospinal tract consists almost entirely of axons that have decussated in the medulla. Most of the corticospinal fibers that do not decussate in the medulla, however, do decussate in the cord before synapsing with neurons in the cord.

______

______

10 Pyramidal tract fibers shown in Fig. 5-8 synapse with spinal cord neurons. The dendrites and cell bodies of pyramidal tract neurons are located in the ....,

and their axons terminate in the ....

---

11 Neurons whose axons comprise the pyramidal tracts, and many other neurons whose cell bodies lie in brain gray matter, are referred to as upper motoneurons. The only neurons called lower motoneurons are those whose axons terminate in skeletal muscles. Because of the location of their cell bodies, lower motoneurons are also called anterior? posterior? horn cells.

---

12 In order for you to voluntarily contract any skeletal muscle, pyramidal tract fibers must conduct impulses from the primary motor area of your cerebral cortex down your cord to stimulate directly, or indirectly (via internuncial neurons), the .... that conduct impulses out to skeletal muscles.

---

13 The ability to voluntarily move a part of the body indicates that these two structures are functioning: the .... tracts from the .... and the anterior horn neurons to the .... muscles.

---

14 If either the pyramidal tracts or the lower motoneurons can function, but not both, voluntary or willed muscle contraction is impossible? is still possible?

---

15 The only way impulses can reach skeletal muscles (somatic effectors) from the spinal cord is via ....

---

16 A relay of at least two? three? neurons conducts motor impulses from the primary motor area to skeletal muscle effectors.

---

17 A relay of at least two? three? neurons conducts sensory impulses from receptors to the general sensory area of the cerebral cortex.

Self-test for Chapters 5 to 7 appears on pp. 123-125.

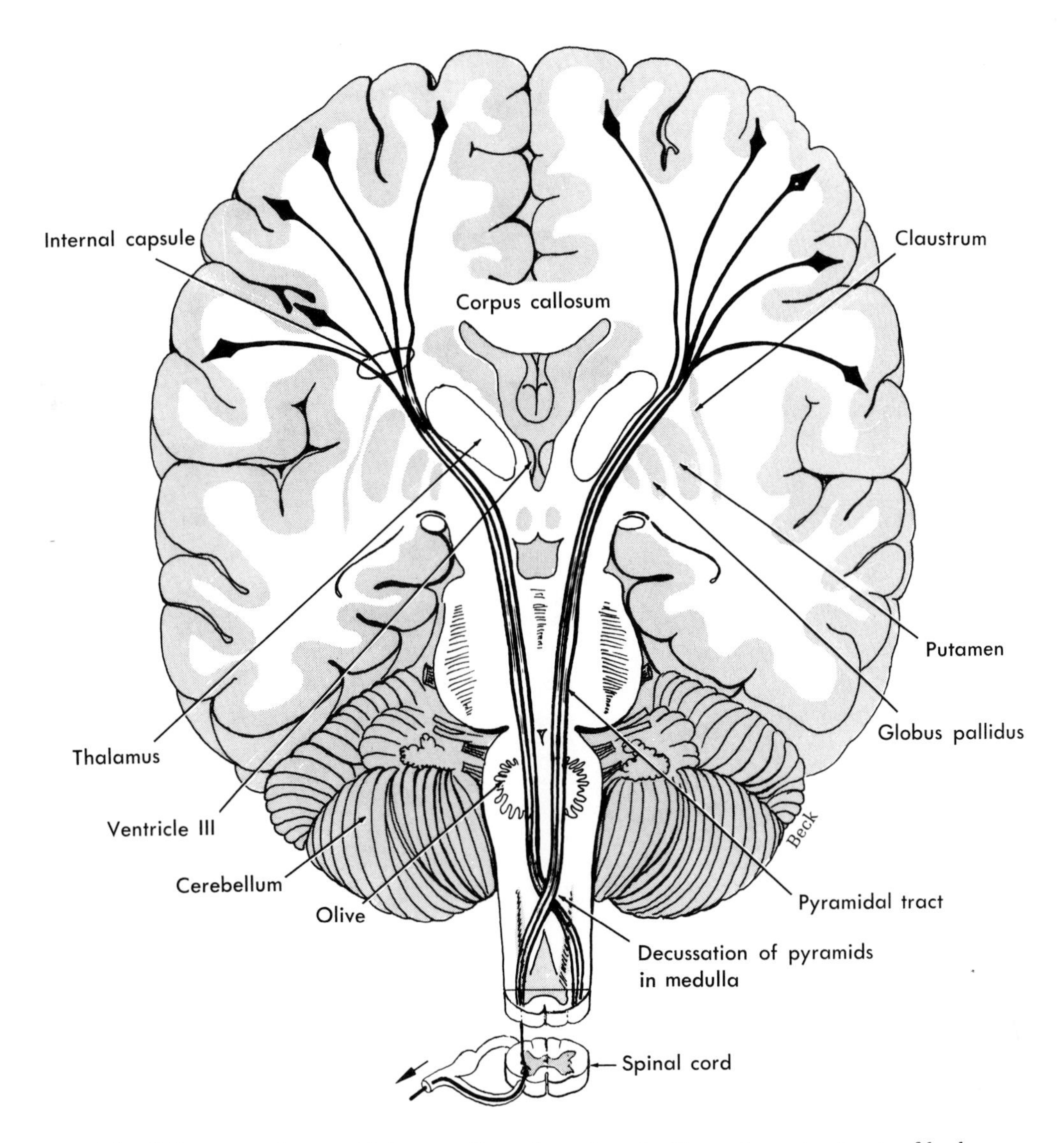

Fig. 5-8 Crossed pyramidal tracts (lateral corticospinal)—main motor tracts of body.

6 The autonomic nervous system

The autonomic nervous system

Procedure A—Investigation of autonomic nervous system

Equipment

1 Charts and illustrations
2 Textbooks on anatomy, physiology, and neuroanatomy

Problems

1 What macroscopic structures compose the autonomic nervous system?
2 From what macroscopic structure to what macroscopic structure does a preganglionic autonomic neuron conduct impulses?
3 From what macroscopic structure to what macroscopic structure does a postganglionic autonomic neuron conduct impulses?
4 Are autonomic neurons sensory or motor or both?
5 What three kinds of tissues constitute autonomic (visceral) effectors?
6 How does the fewest number of neurons necessary to conduct impulses from the spinal cord to somatic effectors compare with the fewest number of neurons necessary to conduct impulses from the cord to autonomic effectors?

Collection of data

1 Examine charts and illustrations.
2 Study a text discussion on the autonomic nervous system.
3 Study Fig. 6-1. Use the data you collect from it to answer questions 1 and 3 to 8 under conclusions.

Conclusions

1 The arrows in Fig. 6-1 indicate that the autonomic neurons shown (like all autonomic neurons) conduct impulses to? from? the cord and therefore are motoneurons? sensory neurons?

____

____

2 If a structure is classified as an autonomic effector, you know that it consists of one of three kinds of tissues—viz., ... , ... , and . . . .

____

____

____

3 As Fig. 6-1 shows, the dendrites and cell bodies of preganglionic (sympathetic) neurons lie in the anterior? lateral? posterior? gray columns of the cord.

____

4 All preganglionic neurons, like the one shown in Fig. 6-1, conduct impulses from the cortex? spinal cord? (or from the brainstem) to autonomic effectors? autonomic ganglia?

____

____

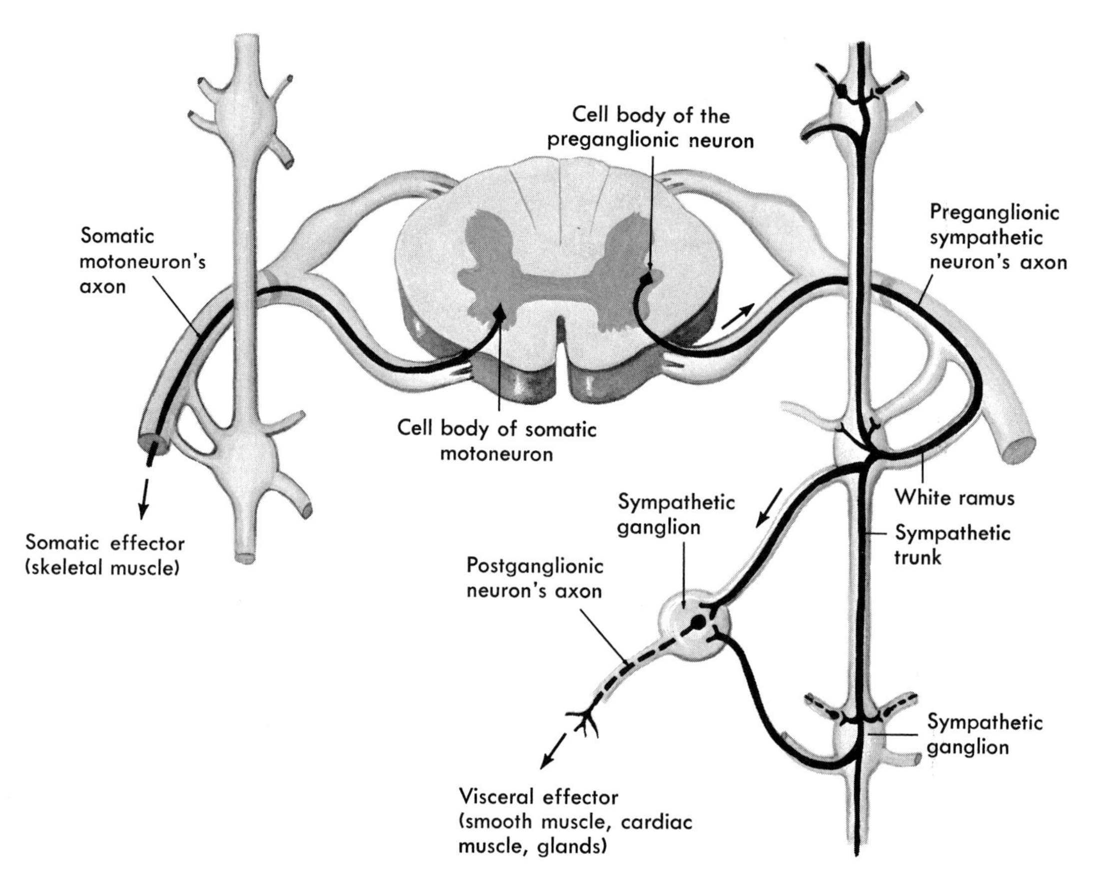

Fig. 6-1 Diagram showing difference between sympathetic pathways from spinal cord to visceral effectors and pathway from cord to somatic effectors.

5 Parasympathetic ganglia consist, as do the sympathetic ganglia shown in Fig. 6-1, mainly of the dendrites and cell bodies of preganglionic? postganglionic? neurons.

____

6 Preganglionic neurons synapse with postganglionic neurons in autonomic (sympathetic or parasympathetic) . . . .

____

7 Whereas preganglionic neurons conduct impulses from the . . . . to autonomic ganglia, postganglionic neurons conduct impulses from the autonomic ganglia to . . . .

____

____

8 The conduction of impulses from the cord to any somatic effector requires one? two? neuron(s), whereas the conduction of impulses to any visceral effector requires . . . . neuron(s).

____

____

9 Whereas any somatic? visceral? effector consists of any one of three kinds of tissues, all . . . . effectors consist of the same kind of tissue, namely, ...

____

____

____

■ The autonomic nervous system—cont'd

Procedure B—Investigation of some autonomic reflexes

Equipment

1 Flashlight
2 Textbook on anatomy and physiology

Problems

1 What is an autonomic reflex?
2 How does the motor pathway in an autonomic reflex arc differ from the motor pathway in a somatic reflex arc?
3 What is the pupillary light reflex?
4 What pair of cranial nerves contains sensory fibers for this reflex arc?
5 What pair of cranial nerves contains motor fibers for this reflex arc?
6 What is the ciliospinal reflex?
7 What pair of cranial nerves contains sensory fibers for this reflex arc?
8 What pair of cranial nerves contains motor fibers for this reflex arc?
9 What is the oculocardiac reflex?
10 What cranial nerve contains sensory fibers for this reflex arc?
11 What cranial nerve contains motor fibers for this reflex arc?

Collection of data

1 Elicit the pupillary light reflex in this way. Ask your laboratory partner to sit with her eyes closed for 1 minute. Then, ask her to open them, and as she does so, flash a bright light into them and note what change in the size of the pupils results.
2 Elicit the ciliospinal reflex in this way. Watch the pupils of your partner's eyes as you pinch her neck firmly enough to cause pain.
3 Elicit the oculocardiac reflex as follows:
   a Count your partner's pulse when she is sitting down. Record results.
   b Use your thumbs to exert firm pressure over your partner's eyeballs. Again count the pulse and record its rate.
4 Consult your textbook to collect data needed for answering problems 4, 5, 7, 8, 10, and 11.

Conclusions

1 Whereas all somatic reflexes consist of .... muscle contractions, autonomic reflexes consist of ...., ..., and ....

____

____

____

____

2 The motor pathways from the cord or brainstem to effectors differ in autonomic and somatic reflex arcs. In autonomic reflex arcs, the motor pathway consists of ...., whereas in somatic arcs it consists of ....

____

____

3 The pupillary light reflex is a constriction? dilatation? of the pupil in response to stimulation of the retina by bright light.

____

4 The visceral effectors that respond in the pupillary light reflex consist of .... of the iris.

____

5 Cranial nerves I? II? IV? contain the sensory fibers for the pupillary light reflex arc.

____

6 Cranial nerves II? III? IV? contain the motor fibers for the pupillary light reflex arc.

____

7 The ciliospinal reflex is a constriction? di-

latation? of the pupil in response to a painful stimulus.

---

8 The oculocardiac reflex is an acceleration? a slowing? of the heart rate in response to pressure over the eyeballs.

---

9 Cranial nerves II? V? X? contain the sensory fibers for the oculocardiac reflex arc.

---

10 Cranial nerves II? V? X? contain the motor fibers for the oculocardiac reflex arc.

Self-test for Chapters 5 to 7 appears on pp. 123-125.

7 Sense organs

■ Sense organs

Procedure A—Investigation of structure of eye

Equipment

1 Fresh beef eyes
2 Beef eyes soaked in 10% formalin overnight
3 Dissecting trays
4 Dissecting instruments

Problem

What gross structural features characterize the eye?

Collection of data

1 Study Fig. 7-1 and examine the horizontal section of an eyeball that has been soaked in formalin overnight. Interpret the data you collect from these by answering the questions under conclusions.
2 Examine the outside of the eyeball to identify the conjunctiva, extrinsic muscles, optic nerve, and fat pads.
3 Cut muscles and fat away from eyeball but leave the optic nerve. Answer question 3 under conclusions.
4 With a scalpel, make an incision about ¼ inch posterior to the edge of the cornea. Insert scissors into the opening and, holding them parallel to the wall of the eyeball, cut completely around the eyeball. Lift off the anterior section. Turn the posterior part of the eyeball upside down over the dissecting tray to allow the lens and vitreous humor to fall out gradually.
5 Place the posterior section of the eyeball in a beaker of water. The retina will probably resume its normal position. Observe the "blind spot"—i.e., the spot on the retina where the optic nerve enters. Note also the retinal blood vessels. No macula lutea can be seen in the beef eye.
6 Remove the retina. Note the iridescent greenish color of the underlying coat in the beef eye.
7 Separate the lens from the vitreous humor. Write the word "and" twice on a piece of paper. Place the lens over one of these words and the vitreous humor over the other. Note any change in the size of either word.
8 Wash under running water the inside of the anterior part of the eye (part lifted off in step 4). Remove as much pigment as possible. Note the cleaned surface carefully. Answer question 10 under conclusions.
9 Hold the lens at eye level and several inches away. Look at your partner. Answer question 14 under conclusions.

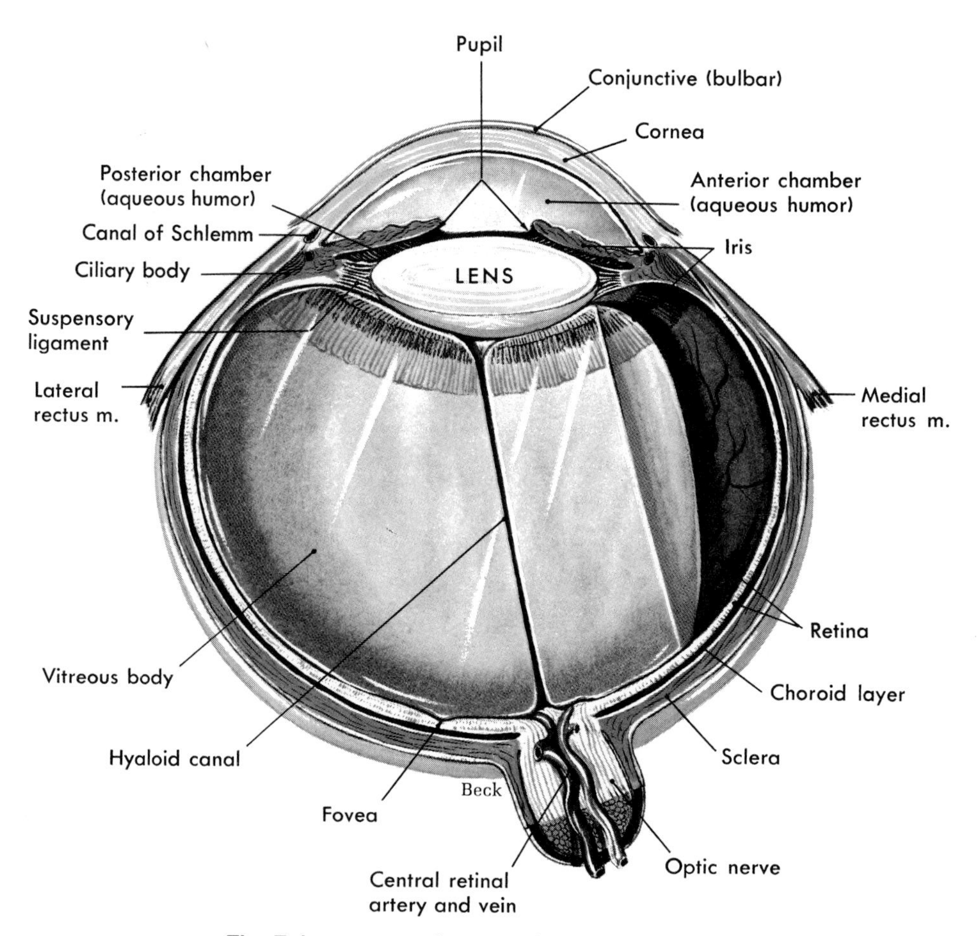

Fig. 7-1 Horizontal section through left eyeball.

Conclusions

1 Conjunctiva is mucous membrane that lines the eyelids and, as Fig. 7-1 shows, it also covers the anterior? posterior? part of the eyeball.

____

2 From the appearance of the muscles that you cut away from the beef eye, you would deduce that they consist of branching? smooth? striated? muscle tissue.

____

3 Removal of muscles and fat from around the eyeball exposes the choroid coat? retina? sclera?

____

4 The anterior cavity of the eyeball is the fluid-filled space anterior to the ...., the suspensory ligament, and the ciliary body. An anterior and a posterior chamber make up the anterior cavity.

____

5 As Fig. 7-1 shows, ... humor fills both chambers of the anterior cavity of the eye.

____

6 As you discovered in step 4 under collection of data, and as Fig. 7-1 shows, a gelatinous substance—viz., the vitreous humor (body)—fills the .... cavity of the eye.

---

7 As you also could observe in step 4, the pupil of the eye is actually a .... in the central part of the iris.

---

8 The blind spot (optic disc) is the place where the .... enters the eyeball.

---

9 The innermost but incomplete coat of the eyeball is the ....

---

10 The structure you removed so much dark pigment from in step 8 under collection of data is the ....

---

11 What kind of tissue composes the iris?

---

12 Do autonomic or somatic motoneurons innervate the iris?

---

13 Do autonomic or somatic motoneurons innervate the external muscles of the eye?

---

14 When you looked at your partner in step 9 under collection of data, did she appear to be right side up or upside down?

■ Sense organs—cont'd

Procedure B—Experiment to demonstrate blind spot

Hold this page about 20 inches from your face with the cross in Fig. 7-2 directly in front of the right eye. You should be able to see both the cross and the circle when you close the left eye. Now, keeping the left eye closed, slowly bring the page closer to the face while fixating the right eye on the cross. At a certain distance, the circle will disappear from your field of vision because its image falls upon the blind spot.

Procedure C—Investigation of structure of ear

Equipment

1 Model of ear
2 Illustrations
3 Textbook of anatomy and physiology

Problem

What gross structural features characterize the ear?

Collection of data

1 Examine a model of the human ear and use Fig. 7-3 to identify its various parts.
2 Interpret your findings by answering the questions listed under conclusions.

From Zoethout, W. D.: Laboratory experiments in physiology, St. Louis, The C. V. Mosby Co.

Fig. 7-2

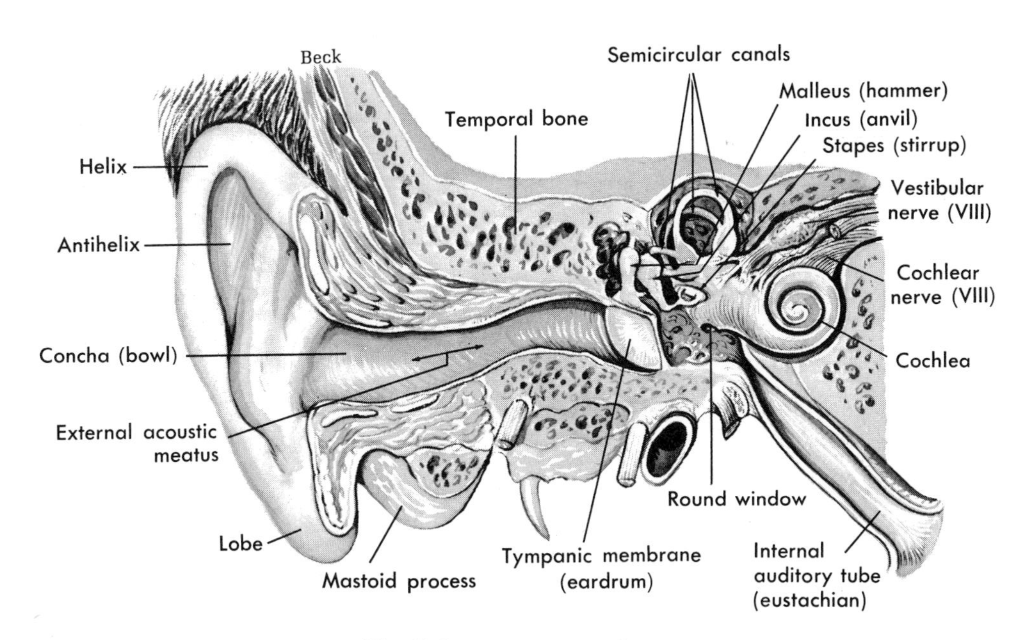

Fig. 7-3 Components of ear.

Conclusions

1 All parts of the ear, except that appended to the head, lie within the .... bone.

2 The anatomical name for the eardrum is ....

3 The sense organ for hearing (viz., the organ of Corti) lies inside a snail-shaped structure—i.e., the .... —located in the inner ear.

4 Because of the location of the organ of Corti, the nerve composed of sensory fibers from this structure is called, as you can see in Fig. 7-3, the .... It is a branch of cranial nerve I? II? VI? VIII?

5 Both the cochlear and the vestibular nerves are branches of the same cranial nerve and both consist of sensory fibers from the inner? middle? ear.

6 The "hearing nerve" is the .... nerve, and the "equilibrium nerve" is the other branch of the eighth cranial nerve—i.e., the .... nerve.

7 There are two equilibrium sense organs—one named the crista ampullaris and located in the semicircular canals and another named the macula located in the vestibule. The semicircular canals, vestibule, and cochlea are the three divisions of the bony labyrinth that constitute the inner? middle? ear.

8 If disease or injury made either the organ of Corti or the cochlear? vestibular? nerve unable to function, you would lose your sense of ....

9 If disease or injury made either the sense organ called the macula or the one called the .... or the cochlear? vestibular? nerve unable to function, you would lose your sense of ....

■ Sense organs—cont'd

Procedure D—Experiments on taste

Equipment

1 Cane sugar solution—8 gm (about 2 tsp) of sugar in 100 ml of water
2 Sodium chloride solution
  Lemon juice
  0.1% quinine solution
3 Small brush or swab
4 Raw potato, onion, apple

Problems

1 Do the same kinds of stimuli activate all taste receptors?
2 What is the pattern of distribution of taste receptors?
3 Does the sense of smell contribute to the sense of taste?
4 Which cranial nerves conduct impulses that result in taste sensations?

Collection of data

1 Dip a small brush or swab in the sugar solution. At intervals of a few seconds, moisten successively the tip, sides, and back of your outstretched tongue and the roof of your mouth. Note results.
2 Repeat with other solutions, rinsing your mouth with clear water after each. Note results with each solution.
3 Ask your partner to close her eyes and pinch her nostrils closed. Then place, in sequence, small, approximately equal-sized pieces of potato, onion, and apple on her tongue. Note whether or not she could quickly identify each piece of food by its taste alone.

Conclusions

1 Data collected in this experiment indicate that different kinds of taste stimuli activate taste buds located in all? different? areas of the tongue.

____

2 Results observed in step 3 under collection of data indicate that the sense of smell does? does not? make a significant contribution to one's sense of taste.

____

3 Which tastes, if any, did you sense in the roof of your mouth?

____

4 Sensations of taste result from impulse conduction by cranial nerves (identify by number and name).

____

____

■ **Self-test** (Chapters 5 to 7)

1 Whereas the inner core of the spinal cord consists of .... matter, the columns around the core consist of ....

2 The small swelling on the posterior root of each spinal nerve is called a ....

3 Cell bodies of sensory neurons that conduct from the periphery to the cord are located in ....

4 Anterior roots of spinal nerves contain axons? dendrites? of sensory neurons? of somatic motoneurons? of postganglionic neurons?

5 Cell bodies of somatic motoneurons are located in the ....

6 Spinal nerves contain axons? dendrites? of sensory neurons and also? but not? axons of motoneurons.

7 The only cell bodies in the anterior gray columns are those of ....

8 Disease or injury that makes posterior root fibers unable to conduct would produce a loss of sensation? paralysis?

9 Effectors are the distal ends of sensory neurons? structures innervated by distal ends of motoneurons?

10 Every effector? receptor? consists of either muscle or glandular epithelial tissue.

11 Stretch reflexes are mediated by two-neuron? three-neuron? reflex arcs.

12 Withdrawal reflexes are mediated by two-neuron? three-neuron? reflex arcs.

13 The reflex centers for both stretch and withdrawal reflexes lie in gray? white? matter of the ....

14 Effectors for both stretch and withdrawal reflexes consist of branching? smooth? striated? muscle.

15 If the spinal nerve that sends branches to the skin and muscles of an area were severed, all but one? all? of the following would characterize that area: loss of reflex movement; loss of sensation; loss of voluntary movement.

16 Most stretch? withdrawal? reflexes are extensor reflexes—i.e., contractions by extensor muscles.

17 Most stretch? withdrawal? reflexes are flexor reflexes.

18 A patient has suffered a subdural hemorrhage—in other words, bleeding under the inner? middle? outer? layer of the ....

19 If both of your cochlear nerves were un-

able to conduct impulses, you would be ....

20 You would be .... if your second cranial nerves could not conduct impulses.

21 If your first cranial nerves could not conduct impulses, you would be unable to ....

22 .... is the term that means a bundle of sensory and? or? motor axons located in the brain or cord.

23 In order for you to sense exactly where something has touched you, none? one? of the following tracts would have to function: lateral corticospinal; lateral spinothalamic; ventral spinothalamic.

24 If the .... tracts are functioning, you would be able to sense heat and cold.

25 If your .... tracts were not able to conduct impulses, you could not feel any pain.

26 If your .... tracts were unable to conduct impulses, your reflexes would still be present, but you could not make any willed movements.

27 Suppose that a physician were to blindfold you and then place a vibrating tuning fork on your arm. If the spinal cord tracts named .... and .... were unable to conduct impulses, you would not sense the vibrations.

28 Another name for pyramidal tracts is .... They are sensory? motor? tracts.

29 The crossed pyramidal tracts are located in the lateral? posterior? ventral? white columns of the cord.

30 All pyramidal tract fibers conduct impulses to .... Some pyramidal fibers synapse with these neurons directly, but most of them synapse with them indirectly via internuncial neurons.

31 The polio virus makes the lower motoneurons that it attacks unable to conduct impulses. This, in turn, makes voluntary contractions and? but not? reflex contractions of the muscles supplied by those motoneurons impossible.

32 The sensory relay to the cerebral cortex consists of a minimum of .... neurons, whereas the motor relay from the cerebral cortex to skeletal muscles consists of .... neurons.

33 Certain motoneurons and? but not? sensory neurons are classified as autonomic neurons.

34 What kind of tissue (or tissues) constitute autonomic effectors?

35 Which of the following is not an autonomic effector: blood vessels (muscle in walls

of)? heart? iris? triceps brachii? salivary glands?

---

36 All? all but one? of the following are autonomic reflexes: dilatation of pupil; heartbeat; eye movements (e.g., to right and left); peristalsis.

---

37 Acetylcholine is the chemical transmitter released by parasympathetic and? but not? sympathetic preganglionic fibers.

---

38 The chemical transmitter released by most sympathetic postganglionic fibers is . . . .

---

39 Presumably all? only some? preganglionic fibers release acetylcholine, as do essentially all parasympathetic? sympathetic? postganglionic fibers.

---

40 Most visceral effectors are innervated by sympathetic and? or? parasympathetic fibers.

---

41 Usually, parasympathetic? sympathetic? impulses dominate the control of most of our visceral effectors.

---

42 All? all but one? of the following are sympathetic responses: dilatation of arterioles in skeletal muscles; dilatation of pupils; increased sweating; slowed peristalsis.

---

43 Visual receptors are located in what structure of the eye?

---

44 Part of the sclera forms the colored? white? part of the eye.

---

45 The pupil of the eye is actually a hole in the . . . .

---

46 The tympanic membrane is the anatomical name of the . . . .

---

47 Destruction of the organ of Corti would produce . . . .

---

48 Destruction of the crista and of the macula would produce loss of the kinesthetic sense? of the sense of equilibrium? of the vibratory sense?

---

49 The crista, macula, and organ of Corti are located deep inside the . . . . bone.

---

50 If you are about to swallow a bitter pill, you will probably taste its bitterness less if you place it on the back? tip? of your tongue.

8 The endocrine system

■ The endocrine system

Procedure A—Film

"Hormonal control of behavior," color, 16 min; McGraw-Hill Films, Dept. WP, 330 West 42nd St., New York, N. Y. 10036 (purchase, $215; rental, $12.50).

Procedure B—Film

"The adrenal," color, 18 min; Department of Anatomy, Duke University School of Medicine, Durham, N. C. 27706.

Procedure C—Investigation of pituitary and adrenal hormones

Equipment

None

Problems

1 What major effects do anterior pituitary hormones produce?
2 What major effects do posterior pituitary hormones produce?
3 What major effects do the hormones secreted by the adrenal cortex produce?
4 What major effects do the hormones secreted by the adrenal medulla produce?

Collection of data

Study the table on pp. 128-129. Then interpret the data you find there by answering the items appearing under conclusions.

Conclusions

1 Several hormones control the metabolism of each of the three kinds of food. Some act as antagonists in their effects; others function as synergists—i.e., they tend to produce opposite? similar? effects.

____

2 The anterior pituitary gland releases into the blood a hormone—viz., ... — that tends to increase protein anabolism. The adrenal cortex secretes some hormones called .... that tend to increase tissue protein breakdown and catabolism.

____

____

3 Because growth hormone, secreted by the ...., accelerates the anabolism of protein, and glucocorticoids accelerate the .... of tissue protein, they are said to be antagonistic? synergistic? in their effects on protein metabolism.

____

____

____

Pituitary and adrenal hormones

Hormone	Functions
Anterior pituitary gland (adenohypophysis) secretes:	
1 Growth hormone (GH)	Promotes protein anabolism (hence essential for normal growth and repair of all tissues)
	Promotes fat mobilization and catabolism—i.e., causes shift from carbohydrate catabolism to fat catabolism
	Slows carbohydrate metabolism; has anti-insulin, hyperglycemic, diabetogenic effect
2 Prolactin (lactogenic hormone)	Promotes breast development during pregnancy
	Initiates milk secretion after delivery of baby
3 Thyroid-stimulating hormone (TSH)	Essential for stimulating thyroid gland growth and secretion
4 Adrenocorticotropic hormone (ACTH)	Essential for stimulating adrenal cortex growth and secretion of glucocorticoids
	Stimulates melanocytes to form pigment melanin
5 Follicle-stimulating hormone (FSH)	Essential for stimulating primary graafian follicle to start growing and to develop to maturity
	Stimulates follicle cells to secrete estrogens
	In male, FSH is essential for stimulating development of seminiferous tubules and spermatogenesis by them
6 Luteinizing hormone (LH)	Stimulates growth of follicle and ovum to maturity
	Stimulates estrogen secretion by follicle
	Causes ovulation; therefore, LH also known as ovulating hormone
	Causes formation of corpus luteum in ruptured follicle following ovulation; hence luteinizing hormone
	In male, LH called ICSH (interstitial cell–stimulating hormone) because it stimulates interstitial cells of testes to secrete testosterone
7 Melanocyte-stimulating hormone (MSH)	Stimulates melanocytes to form melanin; ACTH also has this effect
Posterior pituitary gland (neurohypophysis) secretes:	
1 Antidiuretic hormone (ADH)	Increases water reabsorption by kidney's distal and collecting tubules, thereby producing antidiuresis (less urine volume; name based on this effect)
2 Oxytocin	Stimulates powerful contractions by pregnant uterus; name oxytocin from Greek for "swift childbirth"
	Stimulates milk ejection from alveoli (milk-secreting cells) of lactating breasts into ducts; essential before milk can be removed by suckling

Pituitary and adrenal hormones—cont'd

Hormone	Functions
Adrenal cortex secretes:	
1 Glucocorticoids (GC's)—mainly hydrocortisone and corticosterone	Normal blood level of GC's essential for both normal catabolism and anabolism of proteins, fats, and carbohydrates High blood level of GC's: **a** Accelerates breakdown and catabolism of tissue proteins **b** Accelerates mobilization of fats from adipose cells and their catabolism, whenever glucose catabolism is inadequate to supply cells' energy needs **c** Accelerates gluconeogenesis (conversion of proteins or fats to glucose) by liver cells Normal blood level of GC's essential for normal electrolyte and water metabolism; GC's tend to slightly increase reabsorption of sodium and water and excretion of potassium by kidney tubules Higher than normal blood level of GC's (such as occurs in adaptation phase of stress, for example): **a** Produces anti-inflammatory effect **b** Causes atrophy of lymphatic tissues—tissues that form lymphocytes; decreases number of circulating eosinophils and lymphocytes; some lymphocytes form antibodies; others become plasma cells that also form antibodies; antibodies are protein compounds that react with antigens to produce immunity or, sometimes, allergy **c** Decreases ACTH secretion
2 Mineralocorticoids (MC's)—mainly aldosterone	Greatly increase sodium and water reabsorption and potassium excretion by kidney tubules
Adrenal medulla secretes:	
1 Epinephrine	Increases and prolongs effects of sympathetic stimulation Increases liver glycogenolysis (breakdown of glycogen to glucose)
2 Norepinephrine	Increases and prolongs sympathetic effects, but to different degrees from epinephrine in certain instances

4 Because both glucocorticoids and epinephrine tend to decrease? increase? blood glucose, they are said to act as antagonists? synergists? in their effect on blood sugar concentration.

5 As you can discover in the table on pp. 128-129, glucocorticoids tend to increase tissue protein catabolism and also? but do not? help regulate carbohydrate and fat metabolism.

6 The table reveals that hormones secreted by the adrenal cortex? medulla? cause liver cells to accelerate their breakdown of glycogen to glucose, a process called . . . .

7 Glucocorticoids, hormones secreted by the . . . ., stimulate liver cells to convert protein to glucose, a process called . . . .

8 Both glucocorticoids and epinephrine tend to produce hyperglycemia—i.e., high? low? blood sugar. Glucocorticoids produce their hyperglycemic effect by accelerating the process of liver . . . ., whereas epinephrine produces it by accelerating liver . . . .

9 In addition to their effects on food metabolism, glucocorticoids decrease? increase? immunity and . . . . . inflammation.

10 When the normal body is adapting to stress, the blood concentration of glucocorticoids is higher? lower? than normal.

11 A prolonged higher than normal concentration of glucocorticoids in the blood causes atrophy of? hypertrophy of? no change in? lymphatic tissues.

12 A high? low? blood concentration of glucocorticoids decreases ACTH secretion by the anterior pituitary gland.

13 In general, the more glucocorticoids in blood, the . . . . ACTH secreted.

14 If the blood level of glucocorticoids were chronically higher than normal, sodium and water loss? retention? would result.

15 If the blood level of glucocorticoids were chronically higher than normal, potassium deficiency? excess? would result.

16 If the blood level of glucocorticoids were chronically higher than normal, the concentration of glucose in the blood would probably become higher? lower? than normal.

17 From data in the table on pp. 128-129, you can deduce that if a person took a glucocorticoid preparation over a long period of time, he would most likely have an increased resistance? susceptibility? to infection.

18 From data in the table, it would? would not? seem rational to treat an allergy with glucocorticoids.

____

19 What hormone of what gland stimulates the adrenal cortex to secrete glucocorticoids?

____

____

20 Marked retention of fluid and of potassium? sodium? results from excess secretion of aldosterone by the ....

____

____

21 When you are frightened, your heart beats faster both because of the effect on it of parasympathetic? sympathetic? stimulation and of the hormone, ..., which the .... secretes in increasing amounts at such a time.

____

____

____

22 During a woman's reproductive years, one or more primary graafian follicles usually begin developing each month because of the effect on the ovaries of what hormone secreted by what gland?

____

____

23 On the basis of data in the table, one might describe .... secreted by the .... as the "water-retaining hormone" and .... secreted by the .... as the "salt-and-water–retaining hormone."

____

____

____

____

24 What hormone of what gland is appropriately called the "ovulating hormone"?

____

____

25 It is common practice in obstetrics to give a hormone to the mother immediately after delivery of her baby so as to contract her uterus and minimize blood loss. What hormone of what gland?

____

____

■ Self-test

1 The "master gland" of the endocrine system is the .... gland because it secretes hormones that regulate secretion by other endocrine glands—specifically, by the thyroid gland, adrenal cortex, and the ovaries (or testis).

2 Adenohypophysis? Neurohypophysis? is another name for the posterior pituitary gland.

3 A negative feedback mechanism operates between ACTH secretion by the .... gland and glucocorticoid secretion by the adrenal cortex. Whereas ACTH decreases? increases? glucocorticoid secretion, glucocorticoids decrease? increase? ACTH secretion.

4 A negative feedback mechanism also regulates FSH secretion by the .... gland and estrogen secretion by the ovaries. A high blood concentration of estrogen .....-creases FSH secretion and a high blood concentration of FSH .... estrogen secretion.

5 Growth hormone, secreted by the ...... gland, is appropriately called a "protein ....hormone" because it promotes growth by accelerating the synthesis of tissue proteins.

6 Which, if any, of the following could be correctly called a "protein catabolic hormone": ADH? aldosterone? hydrocortisone?

7 Which of the following is correctly called the "ovulating hormone": ACTH? FSH? LH? oxytocin?

8 Because they tend to accelerate the process called ...., glucocorticoids are classified as hyperglycemic? hypoglycemic? hormones.

9 Because it tends to accelerate the process called ...., epinephrine is classified as a hyperglycemic? hypoglycemic? hormone.

10 ACTH acts indirectly as a .... -glycemic hormone because it stimulates the secretion of ....

11 The .... secretes hormones—viz., ...—that indirectly produce an "immunosuppressive" effect.

12 The hormones named .... are sometimes called anti-inflammatory hormones.

13 Glucocorticoids tend to bring about all? more than one but not all? of the following effects: decreased allergic responses; high blood sugar; less resistance to infection; low lymphocyte count.

14 High blood sodium and low blood potassium could result from both? neither? only one? of these conditions: aldosterone deficit, ADH excess.

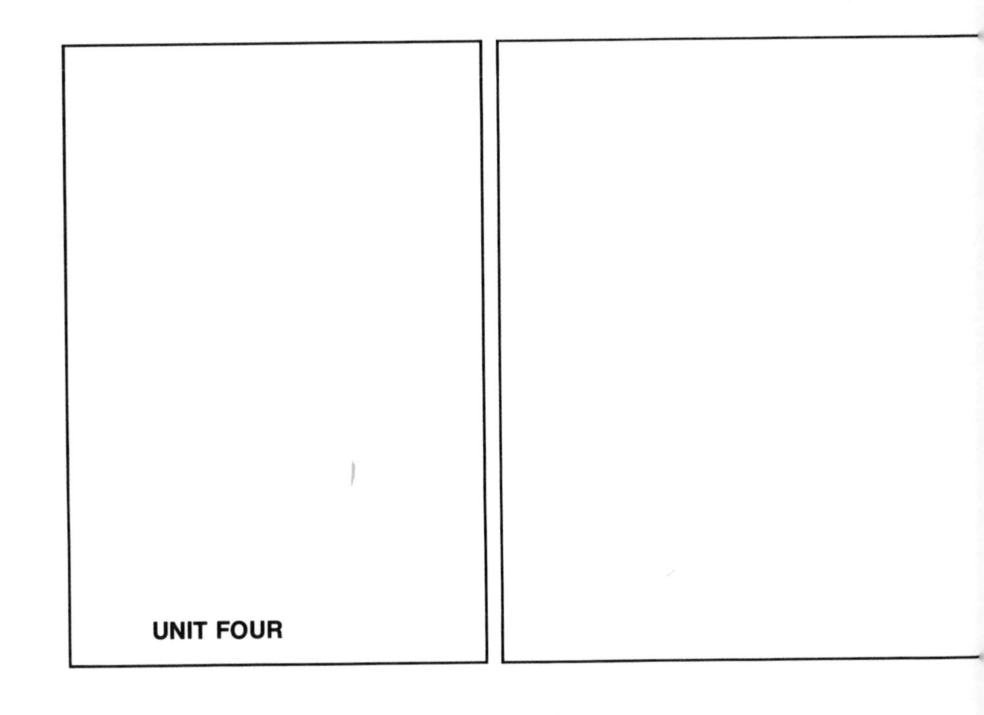

MAINTAINING THE METABOLISM OF THE BODY

To survive, the body must take in certain materials from its environment, use these substances for metabolism, and eliminate various wastes into its environment. This unit consists of four chapters, one for each of the systems most involved with these vital functions. They are the respiratory, circulatory, digestive, and urinary systems.

9 The respiratory system

Anatomy of respiratory system

Procedure A—Film

"Respiratory systems in animals," color, 14 min; Audio-visual Center, Dept. SN, Indiana University, Bloomington, Ind. 47401 (purchase, $210; rental, $8).

Procedure B—Investigation of upper respiratory passages and paranasal sinuses

Equipment

1 Skull
2 Head and torso model
3 Larynx model
4 Charts and illustrations

Problem

What major structural features characterize the upper respiratory passages and nasal sinuses?

Collection of data

1 Examine the skull, models, charts, and illustrations to identify the structures shown in Figs. 9-1 and 9-2.
2 Label Figs. 9-1 and 9-2.

Conclusions

1 On its way to the lungs, air usually first enters the pair of .... cavities. It may, however, enter the .... first.

____

____

2 From the nasal or oral cavities, air must next move through the .... on its way to the lungs.

____

3 In order for air to move into the lungs, it must move out of the pharynx and through the voice-box—i.e., the.... —and on through the ...., the last part of the airway leading to the lungs.

____

____

4 A partition called the .... divides the interior of the nose into two nasal cavities.

____

5 Each nasal cavity consists of three irregularly shaped passageways (meati) because of projections of the .... bone into it.

____

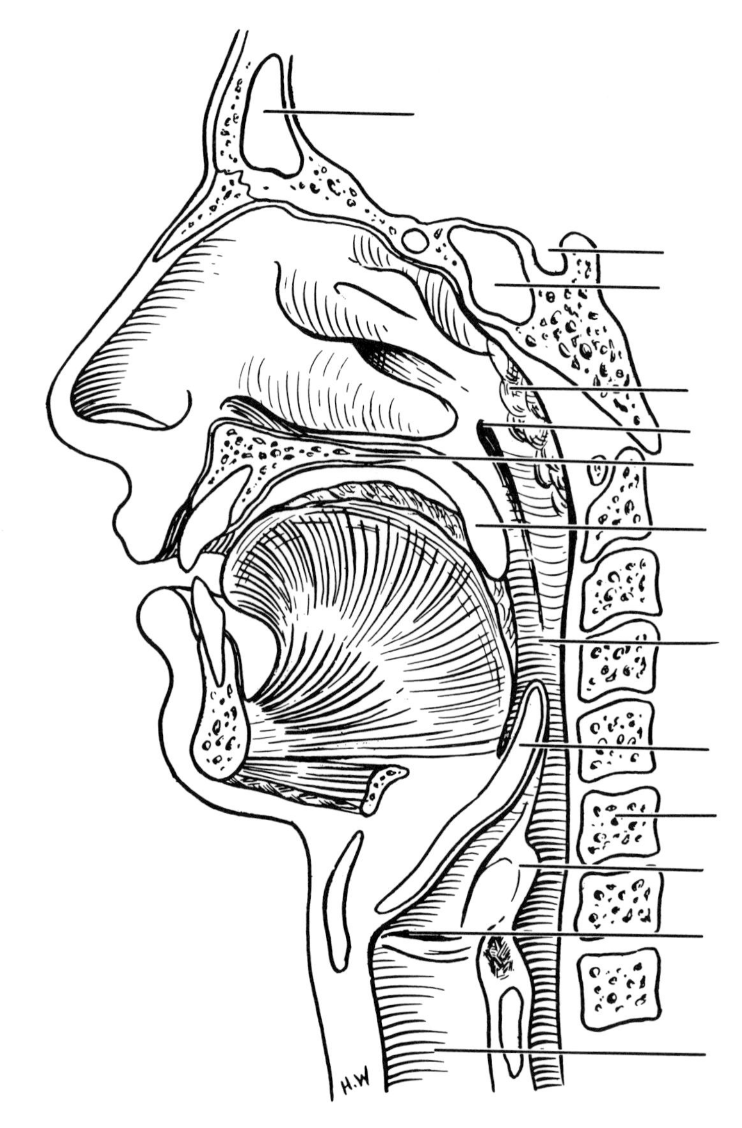

Fig. 9-1 Upper respiratory tract. Place each of the following terms opposite the appropriate label line:

Adenoid
Epiglottis
Frontal sinus
Hard palate (palatine bone)
Larynx
Opening into auditory
  (eustachian) tube
Oropharynx
Sella turcica
Sphenoid sinus
Uvula or soft palate
Vertebra
Vocal fold

Color the respiratory passages red, the digestive passages blue, and the adenoid yellow. Shade the bony sinuses with lead pencil.

Label the hyoid bone.

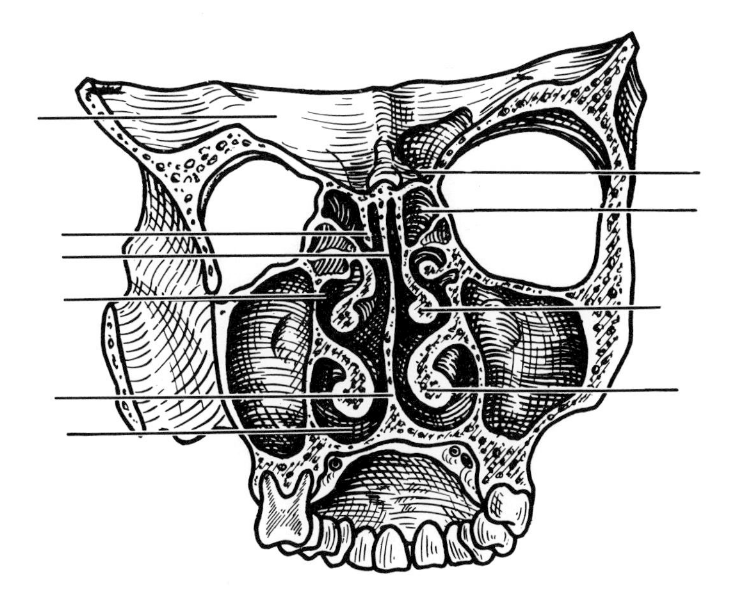

Fig. 9-2 Skull (frontal section). Place each of the following terms opposite the appropriate label line:

Crista galli	Middle concha (turbinate)
Ethmoid sinus	Middle meatus
Frontal bone	Perpendicular plate of ethmoid
Inferior concha (turbinate)	Superior concha (turbinate)
Inferior meatus	Vomer bone

Color the ethmoid bone red and the ethmoid sinuses yellow.

■ Anatomy of respiratory system—cont'd

Procedure C—Dissection of larynx, trachea, bronchi, and lungs of sheep

Equipment

1 Fresh sheep pluck (larynx, trachea, bronchi, lungs, heart, diaphragm, liver); specify to butcher that specimens be left intact, not split open
2 Dissecting instruments
3 Dissecting pans
4 Textbook and chart illustrations

Problem

What main structural features characterize the respiratory system?

Collection of data

1 Study textbook or chart illustrations.
2 Examine the larynx.
  a Note its size, shape, and texture.
  b Note the lidlike structure above the entrance into the larynx. Press on the upper surface of this structure.
  c Feel the shape of the large cartilage that composes most of the anterior wall of the larynx. Note muscles attached to the larynx.
  d Examine the lining of the larynx.
3 Examine the trachea.
  a Observe its size and shape.
  b Try to compress the trachea.
4 Examine the lungs.
  a Try to peel off some of the lung covering.
  b Note divisions of the lungs into lobes and lobes into lobules.
  c Cut off part of the trachea, leaving only 2 or 3 inches of it attached. Insert rubber tubing into the trachea and on into a bronchus. Hold the bronchus closely around the tubing to prevent loss of air as you blow into it. Observe inflation of first one part of the lung and then another as you blow into the tube.
  d Observe the surface of the lung as you stop blowing.
  e Pinch a piece of the lungs between your fingers.
  f Cut out a small piece of the lung and place it in a beaker of water. Note whether or not it sinks.
  g Place a small piece of muscle or liver in a beaker of water. Note whether or not it sinks.
5 Examine a bronchus.
  a Cut open a bronchus and its successively smaller branches.
  b Feel the smallest bronchioles that you are able to cut into.
6 Label Fig. 9-3 as directed.

Conclusions

1 What is the layman's name for the thyroid cartilage?

____

2 What structures shown in Fig. 9-3 are microscopic in size?

____

____

____

3 What structures terminate in these microscopic air sacs?

____

4 The tissue forming the supporting framework of the larynx is ....

____

5 The name of the lidlike structure over the entrance into the larynx is the ....

____

6 Mucous? Serous? Synovial? membrane lines the larynx, trachea, bronchi, etc.

____

7 The trachea does not compress easily because of the presence of C-shaped rings of .... in its wall.

____

8 The name of the serous membrane that forms an adherent covering on the lungs is the ....

---

9 The result observed in steps 4d and 4e under collection of data indicates that lung tissue is? is not? elastic.

---

10 The explanation for results observed in steps 4f and 4g under collection of data is that the lungs contain .... and muscle and liver do not.

---

11 The observations made in step 5 under collection of data indicate that cartilage is? is not? present in the smallest bronchioles.

---

12 All the structures shown in Fig. 9-3 except one serve as air distributors. What one

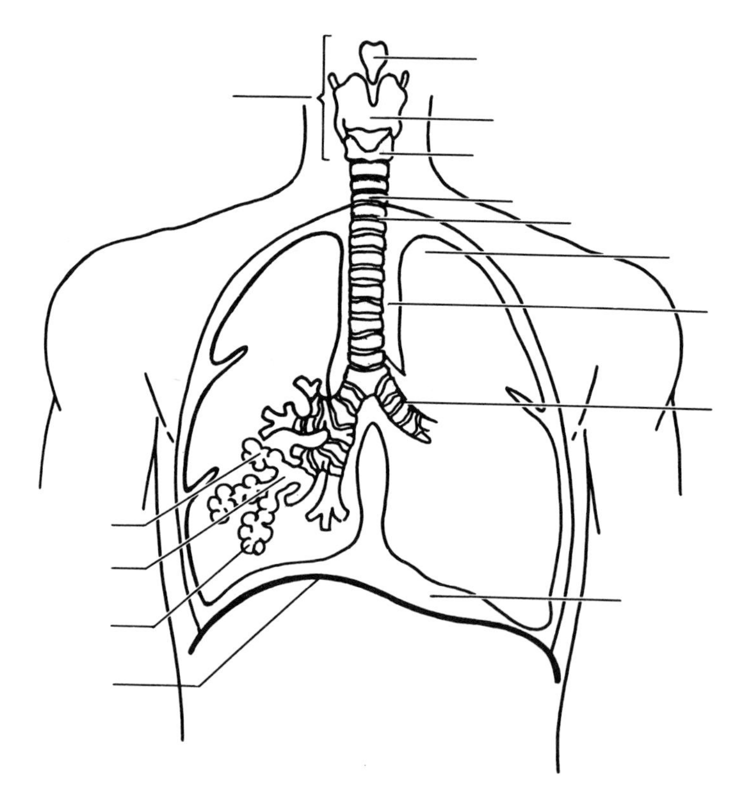

Fig. 9-3 Lungs. Place each of the following terms opposite the appropriate label line:

Alveolar duct
Alveolar sac
Alveolus
Apex of lung
Bronchus
Cricoid cartilage
Diaphragm
Epiglottis
Larynx
Mediastinum
Pleural space
Thyroid cartilage
Trachea

structure that you labeled in this diagram serves as a gas exchanger, the place where air and blood exchange oxygen and carbon dioxide?

____

13 Figs. 9-1 and 9-3 together show the macroscopic structures that function to distribute air to the lungs' millions of alveoli. Name them in order by filling in the following blanks:

a Nasal cavities

b ____________________________

c ____________________________

d ____________________________

e ____________________________

f Bronchioles

14 Since the air-distributing spaces named above do not function in the exchange of respiratory gases, together they are known as the anatomic dead space. Suppose you were to inhale 500 ml air in one breath. Not all of this reaches the alveoli. About 30% or .... ml fills the anatomic dead space. Therefore, not 500 ml of fresh air ventilates the alveoli, but 500 ml minus the dead space volume. In short, the volume of air ventilating the alveoli is always .... than the volume of air inhaled.

■ Anatomy of respiratory system—cont'd
Procedure D—Demonstration of ciliary action

1 Ciliary action can be demonstrated in the following manner:

- **a** Pith both brain and cord of a frog.
- **b** Cut away its lower jaw and slit open the ventral surface of the pharynx and esophagus.
- **c** Lay the frog on its back and hold the slit pharynx and esophagus open with pins.
- **d** Put a piece of cork as small as the head of a common pin on the mucosa of the pharynx.
- **e** Observe it for several minutes.

2 Raise the end of the board or pan on which the frog's head lies to discover whether ciliary action is forceful enough to move the cork "uphill."

Conclusion

What function do you think the ciliated surface of the respiratory mucosa serves?

____

____

■ Physiology of respiratory system

Procedure A—Mechanics of breathing (pulmonary ventilation)

Equipment

1. Textbooks on physiology and on physics
2. Tape measure

Problems

1. How is the thoracic cavity made longer? Broader? Thicker from front to back?
2. How is the change in the size of the thoracic cavity related to inspiration and expiration?
3. What is the relationship between the volume of space occupied by a gas and its pressure when its temperature remains constant?

Collection of data

1. With your partner standing with her profile toward you, observe the anteroposterior diameter of her chest as she takes a deep breath.
2. Have your partner stand with her back toward you. Observe the horizontal diameter of her thorax as she takes the deepest breath possible.
3. Measure and record the chest circumference:
   - **a** At the level of the axillae at the end of a normal expiration
   - **b** At the end of a normal inspiration
   - **c** At the end of the deepest possible inspiration
4. Look up Boyle's law in a textbook on physics if you do not already know it.
5. Formula for converting mm Hg to cm H_2O pressure: Multiply mm Hg by 1.36.

Conclusions

1. Contraction of the . . . . muscle makes the thoracic cavity longer.

___

2. Contraction of the chest-elevating muscles makes the thoracic cavity larger in which dimension(s)?

___

3. Inspiration occurs as the size of the thoracic cavity decreases? increases?

___

4. The thoracic cavity has to decrease in size in order for inspiration? expiration? to occur.

___

5. According to Boyle's law, the pressure exerted by a gas decreases as the volume of space it occupies decreases? increases? provided its temperature remains constant.

___

6. Inspiration is produced by a decrease? an increase? in the pressure in the thoracic cavity and in the air passages of the lungs.

___

7. Under standard conditions, the atmosphere exerts a pressure of . . . . mm Hg.

___

8. Judging from the figures given in Fig. 9-4, how does normal intrathoracic pressure compare with atmospheric pressure at the end of expiration and beginning of inspiration?

___

9. How does normal intrathoracic pressure compare with atmospheric pressure at the end of inspiration and beginning of expiration (judging from the figures in Fig. 9-4)?

___

10. An important fact about intrathoracic pressure is that normally it is always . . . . than atmospheric pressure. In other words, intrathoracic pressure, in relation to atmospheric pres-

sure, is always negative? positive? pressure.

____

____

11 At the end of expiration and beginning of inspiration (according to Fig. 9-4), intrathoracic pressure is .... mm Hg, which is .... cm H_2O pres less than standard atmospheric pressure.

____

12 During normal quiet inspiration, according to Fig. 9-4, intrathoracic pressure .... -creases by .... mm Hg, which is .... cm H_2O pres, whereas during normal expiration it .... by the same amount.

____

____

____

____

13 Explain the mechanism of inspiration by encircling the appropriate terms in the diagram beneath Fig. 9-4.

14 According to Fig. 9-4, the pressure in the alveoli of the lungs at the end of one expiration and before the next inspiration begins is .... -er than? the same as? atmospheric pressure.

____

15 Alveoli can best be thought of as functioning as air distributors? gas exchangers?

____

16 Unless intra-alveolar pressure changes as shown in Fig. 9-4, air cannot move into the alveoli of the lungs and oxygen and carbon dioxide cannot be exchanged between the air and blood. Specifically, alveolar pressure must ..... -crease to a level .... -er than .... pressure in order to cause inspiration.

____

____

____

17 Air can be expired from the lungs only if alveolar pressure ..... -creases to a .... -er level than ....

____

____

____

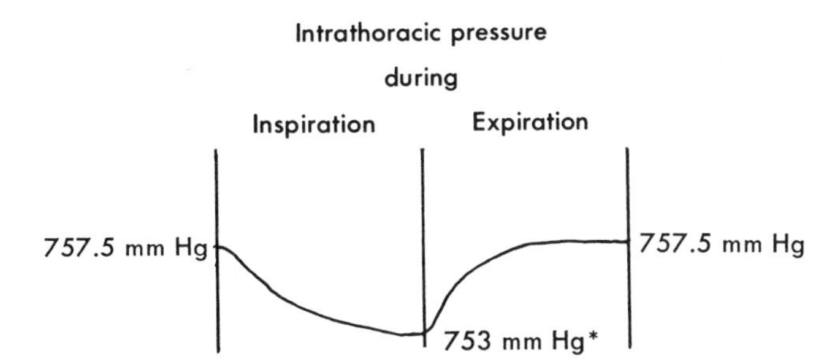

*Normal value for usual (not deep) breathing.

Fig. 9-4 Intrathoracic and intra-alveolar pressures during quiet breathing.

• • •

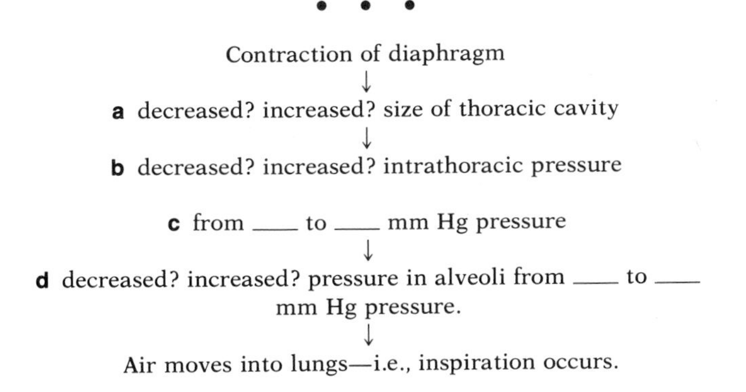

Contraction of diaphragm

↓

a decreased? increased? size of thoracic cavity

↓

b decreased? increased? intrathoracic pressure

c from ____ to ____ mm Hg pressure

↓

d decreased? increased? pressure in alveoli from ____ to ____ mm Hg pressure.

↓

Air moves into lungs—i.e., inspiration occurs.

■ Physiology of respiratory system—cont'd

Procedure B—Experiment to discover relationship between carbon dioxide content of blood and rate and depth of respirations

Equipment

1 Watch with second hand
2 Paper bag

Problem

Is an increase in blood carbon dioxide content followed by changes in the rate and depth of respirations? If so, what changes?

Collection of data

1 Count and record your partner's respirations for 2 minutes as she sits quietly. Note whether they are shallow or deep or in between.
2 Ask your partner to hyperventilate by breathing deeply and rapidly for 2 minutes. Immediately afterward, again count respirations and note their depth for 1 minute. Compare with the results obtained in the preceding step. Also observe whether or not the subject took a breath immediately following hyperventilation or whether there was a period of apnea. Answer questions 1 and 2 under conclusions.
3 This time have the subject hyperventilate 2 minutes into a paper bag held tightly over her mouth and nose. Immediately after, count and record her respirations for 1 minute and note their depth. Compare with the results obtained in the preceding step. Observe also whether the subject took a breath immediately after hyperventilating into the paper bag or whether it was followed by a period of apnea. Answer questions 3 and 4 under conclusions.
4 Wait 2 or 3 minutes. Then ask the subject to pinch her nose shut and hold her breath as long as she can. Note the length of time.
5 Wait 2 or 3 minutes. Then ask the subject to hyperventilate for 2 minutes and then hold her breath as long as possible. Note the length of time. Compare with the results obtained in step 4. Answer questions 5 and 6 under conclusions.
6 a Have the subject run in place for 2 or 3 minutes and then hold her breath as long as possible. Note length of time.
  b Immediately count respirations and note their depth. Compare with the results obtained in step 5. Answer question 7 under conclusions.

Conclusions

1 When a normal individual hyperventilates in fresh air, his blood carbon dioxide content necessarily decreases? increases?

____

2 Results observed in step 2 under collection of data indicate that a decrease in blood carbon dioxide is followed by a decrease? an increase? in the rate and depth of respirations.

____

3 Hyperventilation into the paper bag (rebreathing exhaled air) causes the blood carbon dioxide content to decrease? increase?

____

4 Results observed in step 3 under collection of data indicate that an increase in blood carbon dioxide is followed by a decrease? an increase? in the rate and depth of respirations.

____

5 The subject was able to hold her breath longer when she did? did not? first hyperventilate.

____

6 The explanation for the result noted in the preceding answer is that hyperventilation decreases the .... content of blood

so it takes longer for it to accumulate to the level necessary to stimulate respirations.

---

7 Based on the results observed in these experiments, the rate and depth of respirations are directly? inversely? related to the amount of carbon dioxide in the blood.

---

8 Here is an important principle to remember: normally when blood carbon dioxide content increases, respirations decrease? also increase?

---

9 The corollary of the principle stated in the preceding item is this: when blood carbon dioxide content decreases, respirations also decrease? increase?

■ Physiology of respiratory system—cont'd

Procedure C—Respiratory volumes and capacities

Equipment

Spirometer

Problems

1. How much air do you usually breathe in and out of your lungs?
2. How much air can you force into your lungs after you have inspired your usual normal amount?
3. After you have inspired and expired your usual amount of air, how much more air can you forcibly expire?
4. What is the largest volume of air you can force out of your lungs after you have inspired the largest amount of air you can?

Collection of data

1. Set the pointer on the spirometer at 0 mm.
2. Measure your tidal volume as follows. Take a normal breath, then put the mouthpiece of the spirometer in your mouth and exhale normally into it. Record the number of millimeters the spirometer bell rose. Calculate how many milliliters of air you exhaled by multiplying the number of millimeters the bell rose by the constant K for the spirometer. (K is the volume in milliliters per millimeter rise of the spirometer bell.) Repeat steps 1 and 2 a few times.
3. Repeat step 1. Take the deepest inspiration you can, then forcibly exhale into the spirometer mouthpiece as much air as you can. Note the millimeter rise of the spirometer bell. Calculate volume of air exhaled as in step 2.
4. Repeat step 1. Take a normal breath in, breathe out normally, and then exhale into the spirometer mouthpiece as much air as you possibly can. Note the millimeter rise of the spirometer bell. Calculate the volume of air you forcibly expired after your normal inspiration and expiration. This is your expiratory reserve volume.
5. Repeat step 1. To measure your inspiratory reserve volume, inhale as deeply as you can, then exhale normally into the spirometer. Note the millimeter rise of the spirometer bell. Calculate your inspiratory reserve volume.
6. Examine Fig. 9-5 carefully.

Conclusions

1. According to Fig. 9-5, an example of a normal
   a Vital capacity is .... ml

   ____

   b Tidal volume is .... ml

   ____

   c Inspiratory reserve volume is .... ml

   ____

   d Expiratory reserve volume is .... ml

   ____

2. The volume of air that an individual inhales and exhales when he is breathing quietly is called his ....

   ____

3. The greatest amount of air that an individual can inhale and exhale is called his ....

   ____

4. After a normal inspiration and normal expiration, the additional volume of air that an individual can force out of his lungs is called his ....

   ____

5. From Fig. 9-5, do you conclude that anyone can forcibly expire all the air from his lungs?

   ____

6. The inspiratory capacity of the lungs—i.e., the maximum amount of air that can

be inspired after a normal expiration—changes in certain lung disorders. The inspiratory capacity is the same as the inspiratory reserve volume? sum of the inspiratory reserve volume plus the tidal volume? What is your inspiratory capacity?

____

____

7 A change in the functional residual volume of an individual's lungs may indicate lung disease. The functional residual volume is the amount of air left in the lungs at the end of a normal expiration. From this definition and from Fig. 9-5, you can deduce that the functional residual capacity is the sum of the residual volume plus the ....
What is your functional residual capacity? According to Fig. 9-5, what is it normally?

____

____

____

8 Fig. 9-5 indicates that the total lung capacity is the sum of ....
What is your total lung capacity (using the residual volume shown in the figure)?

____

____

9 If you have a normal tidal volume of 500 ml and a normal anatomic dead space of 150 ml, your alveolar ventilation volume is .... ml. If you were to breathe through a tube with a volume of 150 ml, the volume of your anatomic dead space would then be .... ml, and your alveolar ventilation volume would be .... ml. An important principle is that any condition that increases the anatomic dead space—as certain lung diseases do—........ -creases the alveolar ventilation volume and thereby leads to inadequate amounts of oxygen being added to blood and too little carbon dioxide being removed from it.

____

____

____

____

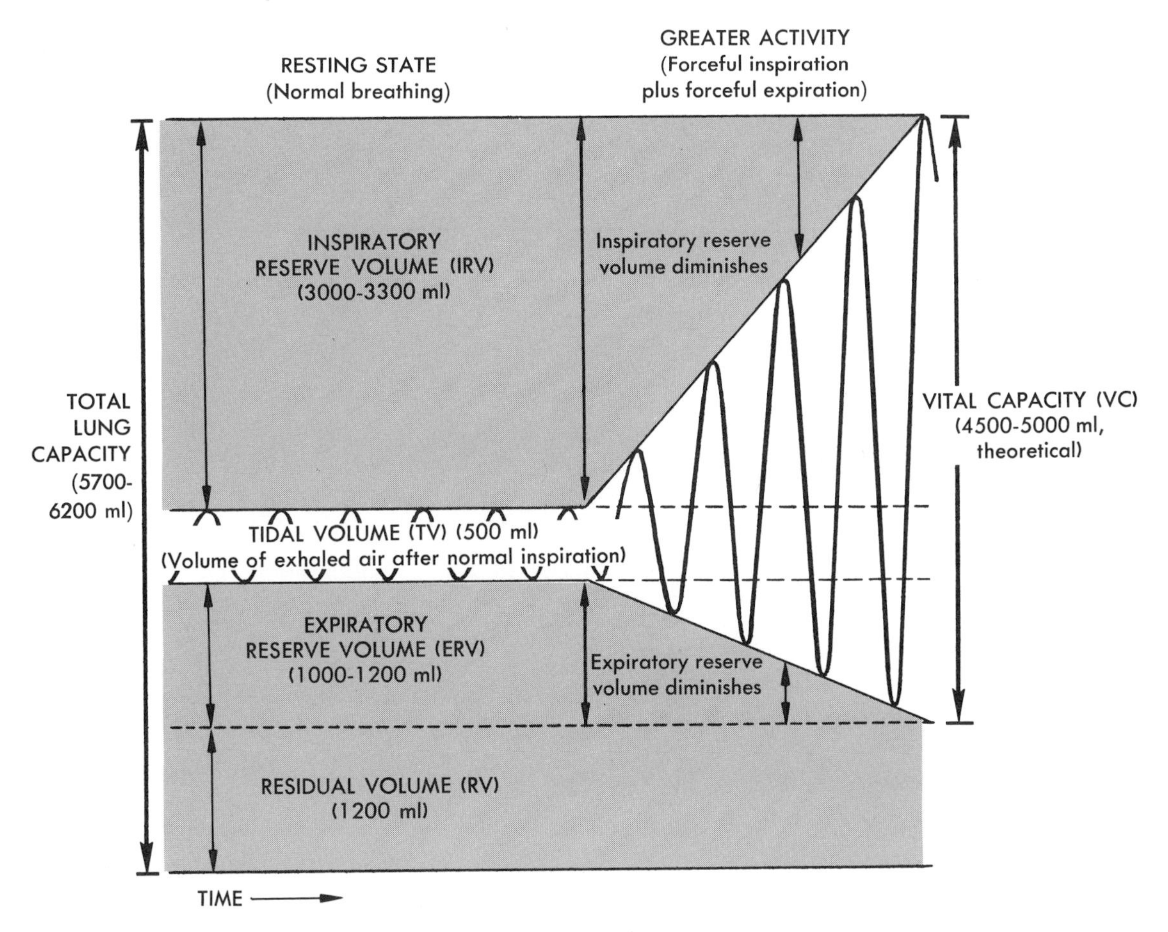

Fig. 9-5 During normal, quiet respirations, the atmosphere and lungs exchange about 500 ml of air (tidal air). With a forcible inspiration, about 3,300 ml more air can be inhaled (inspiratory reserve volume). After a normal inspiration and normal expiration, approximately 1,000 ml of air can be forcibly expired (expiratory reserve volume). Vital capacity is the amount of air that can be forcibly expired after a maximum inspiration and indicates, therefore, the largest amount of air that can enter and leave the lungs during respiration. Residual air is that which remains trapped in the alveoli.

10 The cardiovascular system

■ Functions

Procedure—Film

"The human body: circulatory system," color, 13½ min; Modern Film Rentals, 2323 New Hyde Park Road, New Hyde Park, N. Y. 11040 (rental, $8).

■ Blood

Procedure A—Microscopic examination of blood

Equipment

1 Cotton
2 Glass slides
3 Cover slips
4 Filter paper
5 Needle
6 70% alcohol
7 Wright's stain
8 Distilled water
9 Microscopes

Problems

1 What kinds of cells compose blood?
2 What structural features characterize these cells?

Collection of data

1 After scrubbing your hands, sponge the tip of the left little finger with 70% alcohol and let it dry in air.
2 Puncture with a sterile lancet or needle. Touch a clean glass slide to the blood. Cover with a cover glass.
3 Examine the slide under low power and high power. Note shape, color, and size of the red blood cells.
4 Make a thin blood smear by placing a drop of blood about 1 inch from the end of the slide and smearing it with another slide held on end and at an oblique angle.
5 Stain with Wright's stain:
   a Cover the smear with Wright's stain. Let it stand 1 minute.
   b Add small amount of distilled water. Let it stand 2 minutes.
   c Keep slide horizontal and wash it with distilled water until the smear is pink and translucent.
   d Blot it dry with filter paper.
6 Examine the smear under oil-immersion objective. Use colored illustrations of blood cells as a guide in identifying neutrophils, eosinophils, basophils, lymphocytes, and monocytes.

■ Blood—cont'd

Procedure B—Red blood cell count, white blood cell count, and hemoglobin estimation

Arrange for students to visit a hospital laboratory to observe how red blood cell counts, white blood cell counts, and hemoglobin estimations are done.

Procedure C—ABO and Rh blood typing

Equipment

1. 70% alcohol
2. Cotton
3. Sterile lancet or needle
4. Microscope slides
5. Wax pencil
6. Toothpicks
7. Anti-A, anti-B, and anti-Rh_0 (anti-D) typing sera

Problems

1. What type blood do you have—A? B? AB? O?
2. Do you have Rh-positive or Rh-négative blood?

Collection of data

1. With a wax pencil, label the left half of a microscope slide anti-A and the right half anti-B.
2. Place a drop of anti-A serum on the left half of the slide and a drop of anti-B serum on the right half of it.
3. Scrub your hands, then clean the tip of your left little finger—or of your right finger if you are left-handed—with an alcohol-moistened piece of cotton. Let dry in the air.
4. With a sterile lancet or needle (previously flamed and cooled), prick the tip of your cleaned finger. Place a drop of blood from it into each of the drops of anti-sera on the glass slide. Mix each with a clean toothpick. Watch both mixtures for clumping (agglutination) of red blood cells.
5. Warm a microscope slide to 40° to 45° C. Quickly place 2 large drops of blood and 1 drop of anti-Rh_0 typing serum on the slide. Mix with a toothpick and spread to cover about one fourth of the slide. Observe for 2 minutes for agglutination.
6. Consult a textbook for the meaning of ABO blood types and Rh-positive and Rh-negative blood types.

Conclusions

1. What type blood do you have if your red cells were clumped by the
   a. Anti-A serum only?

      ____

   b. Anti-B serum only?

      ____

   c. Both anti-A and anti-B sera?

      ____

   d. Neither anti-A nor anti-B serum?

      ____

2. Do you have Rh-positive blood or Rh-negative?

   ____

3. What result justified your answer to the preceding question?

   ____

■ Blood—cont'd

Procedure D—Bleeding time

Equipment

1 70% alcohol
2 Cotton
3 Absorbent paper (e.g., facial tissues)
4 Sterile lancet or needle

Problem

How long does bleeding continue from a pricked finger?

Collection of data

1 After your laboratory partner has scrubbed her hands, clean the tip of one of her fingers with an alcohol-moistened piece of cotton. Let it dry in the air.
2 With a sterile lancet or needle, prick her clean fingertip. Note the time at which you do this.
3 Blot the pricked point with a clean facial tissue at intervals until bleeding stops. Note the time at which this occurs.
4 Consult a textbook to check the factors involved in stopping bleeding and to find out the normal bleeding time.

Conclusion

1 How long did it take for the bleeding to stop?

____

2 What is the normal bleeding time?

____

3 What two processes act to stop bleeding?

____

____

4 A deficiency in the number of what kind of blood cells is one cause of a longer than normal bleeding time?

____

Procedure E—Film

"The embattled cell," color, 21 min; American Cancer Society, local office (loan). Time-lapse photography of living cancer cell activity in body and defense by lymphocytes.

Heart

Procedure A—Dissection of animal heart

Equipment

1 Fresh sheep or calf hearts—ask slaughter-house to leave hearts intact in their sacs
2 Dissecting trays
3 Dissecting instruments

Problems

1 What are the gross structural characteristics of a mammalian heart?
2 What function do heart valves serve?

Collection of data

1 Study an illustration showing a frontal section of the human heart in your textbook or some other source book. Then follow the directions given in item 1 under conclusions.
2 Examine the sac enclosing the heart. Note the texture and strength of the tissue that forms the sac. Observe where the sac attaches to the heart. Note, too, whether the heart can move freely in its sac.
3 Cut open the sac and examine its lining. Note that an extension of this same lining membrane adheres to the outer surface of the heart as a covering. Answer questions 2 to 5 under conclusions.
4 Note the large vessels emerging from the upper surface of the heart. Observe the shape of the heart with its apex directed downward. Note the color and texture of the cardiac muscle.
5 Press the wall of the lower part of the heart between your thumb and forefinger. The part that feels noticeably thicker is the left ventricle. Compare thickness of the walls of the upper part of the heart (atria) with thickness of the walls of the ventricles.
6 Find collapsed, thin-walled vessels entering the right atrium. Cut through these into the atrium, making the opening wide enough so that you can look down upon the atrioventricular orifice. Pour water through the orifice into the ventricle. Watch the valve close as the ventricle fills.
7 Empty water out of the right ventricle and cut through the ventricular wall. Examine parts of the valve. Note the number of flaps or cusps, as well as the cords that anchor them to the ventricular wall. Note whether this atrioventricular valve would permit blood to flow down into or up out of the right ventricle.
8 Feel the lining of the heart.
9 Run your finger up along the septum of the right ventricle until it enters a blood vessel.
10 Open this vessel with your scissors. Examine the valve now visible. Note in which direction it permits blood flow.
11 Cut open the left atrium. Find the openings of the four pulmonary veins.
12 Cut down through the left ventricular wall. Examine the valve just below the atrioventricular orifice. Note how many cusps the valve has and whether it permits blood to flow down into or up out of the left ventricle.
13 Run a finger up along the septum of the left ventricle until it enters a blood vessel.
14 Cut open this vessel.
15 Find two small openings behind the flaps of this valve. Insert a blunt probe into first one and then the other of these openings to discover where they lead.

Conclusions

1 Label Fig. 10-1 as directed.
2 The name of the sac that encloses the heart is the . . . .

___

3 The structure and mode of attachment of the pericardium do? do not? facilitate free movement of the heart.

___

4 Mucous? Serous? Synovial? membrane lines the fibrous pericardium and covers the outer surface of the heart.

5 The membranous covering adherent to the outer surface of the heart is named the . . . .

6 The ventricles have much thicker? thinner? walls than do the atria.

7 The chamber of the heart that has the thickest wall of all is the . . . .

8 The thin-walled vessels through which blood enters the right atrium are named the . . . . and . . . .

9 Semilunar valves are found at the beginning of what two vessels?

10 The tricuspid valve allows blood to move from the . . . . into the . . . . but prevents its moving back in the opposite direction.

11 The atrioventricular valve that has only two cusps is named the bicuspid or . . . . . . valve.

12 The vessel emerging from the upper part of the right ventricle is named the . . . .

13 The name of the vessel through which blood leaves the left ventricle is the . . . . .

14 The famous . . . . arteries have their origin behind the flaps of the aortic semilunar valve.

Heart—cont'd
Procedure B—Film

"The human heart," color, 26 min; McGraw-Hill Films, 330 West 42nd St., New York, N. Y. 10036 (purchase, $325; rental, $18).

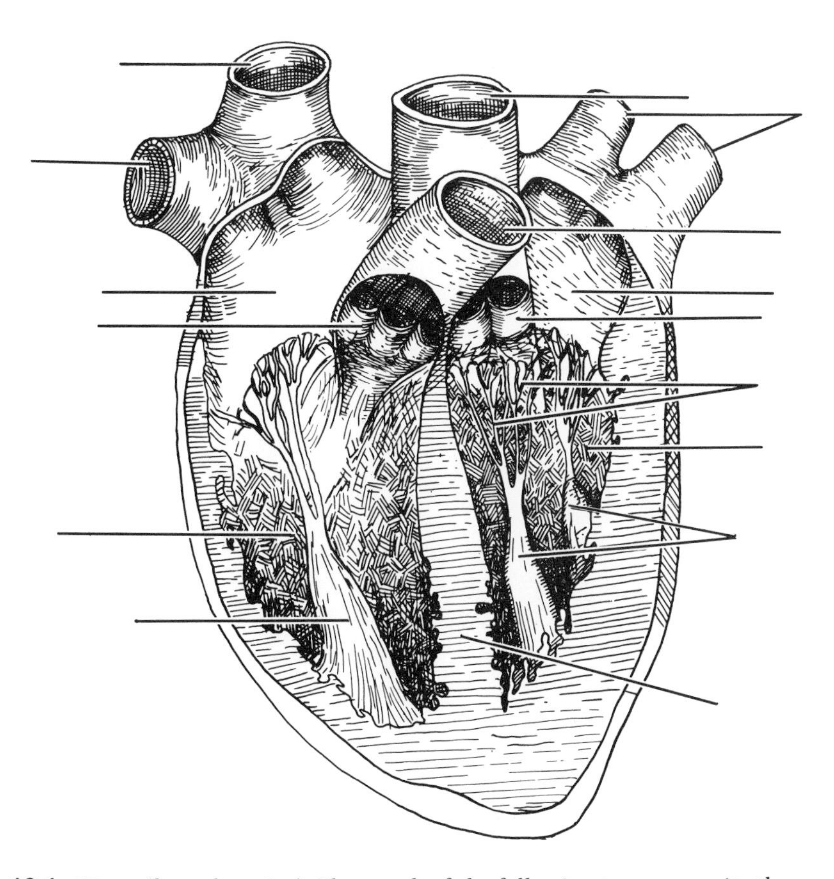

Fig. 10-1 Heart (frontal section). Place each of the following terms opposite the appropriate label line:

Aortic semilunar valve
Ascending aorta
Chordae tendineae
Inferior vena cava
Left atrium
Left ventricle
Papillary muscles
Pulmonary artery
Pulmonary artery semilunar valve
Pulmonary veins
Right atrium
Right ventricle
Septum
Superior vena cava

Color structures that contain venous (deoxygenated) blood blue, those that contain arterial (oxygenated) blood red.

Use blue arrows to show the direction venous blood flows through the heart and red arrows for the direction of arterial blood flow.

Add sinoatrial node, atrioventricular node, and bundle of His in yellow.

Trace the flow of blood through the heart, naming the orifices and the valves of the heart.

■ Blood vessels

Procedure A—Identification of main arteries of body

Equipment

1 Charts and illustrations
2 Torso model of the human body
3 Brain model
4 Textbook

Problem

What are the names and locations of the main arteries of the body?

Collection of data

1 Examine textbook and chart illustrations and the torso model of the human body to identify the following arteries by name and location:
Anterior tibial
Aorta
Arch of aorta
Axillary
Brachial
Celiac
Common carotid
Common iliac
External carotid
External iliac
Femoral
Inferior mesenteric
Innominate
Internal carotid
Internal iliac (hypogastric)
Left coronary
Palmar arch, deep
Palmar arch, superficial
Posterior tibial
Popliteal
Pulmonary
Radial
Renal
Right coronary
Splenic
Subclavian
Superior mesenteric
Ulnar

2 Locate each of the arteries identified in step 1 as nearly as possible on your own body.
3 Examine the charts, illustrations, and models to identify names and locations of the main arteries to the brain.

Conclusions

1 Place the names of the arteries identified in step 1 under collection of data on their appropriate label lines on Fig. 10-2, leaving the other label lines blank.
2 Label Fig. 10-3 as directed.
3 What arteries form the "circle" of Willis?

____

____

____

____

4 Follow the directions given at the bottom of p. 169.

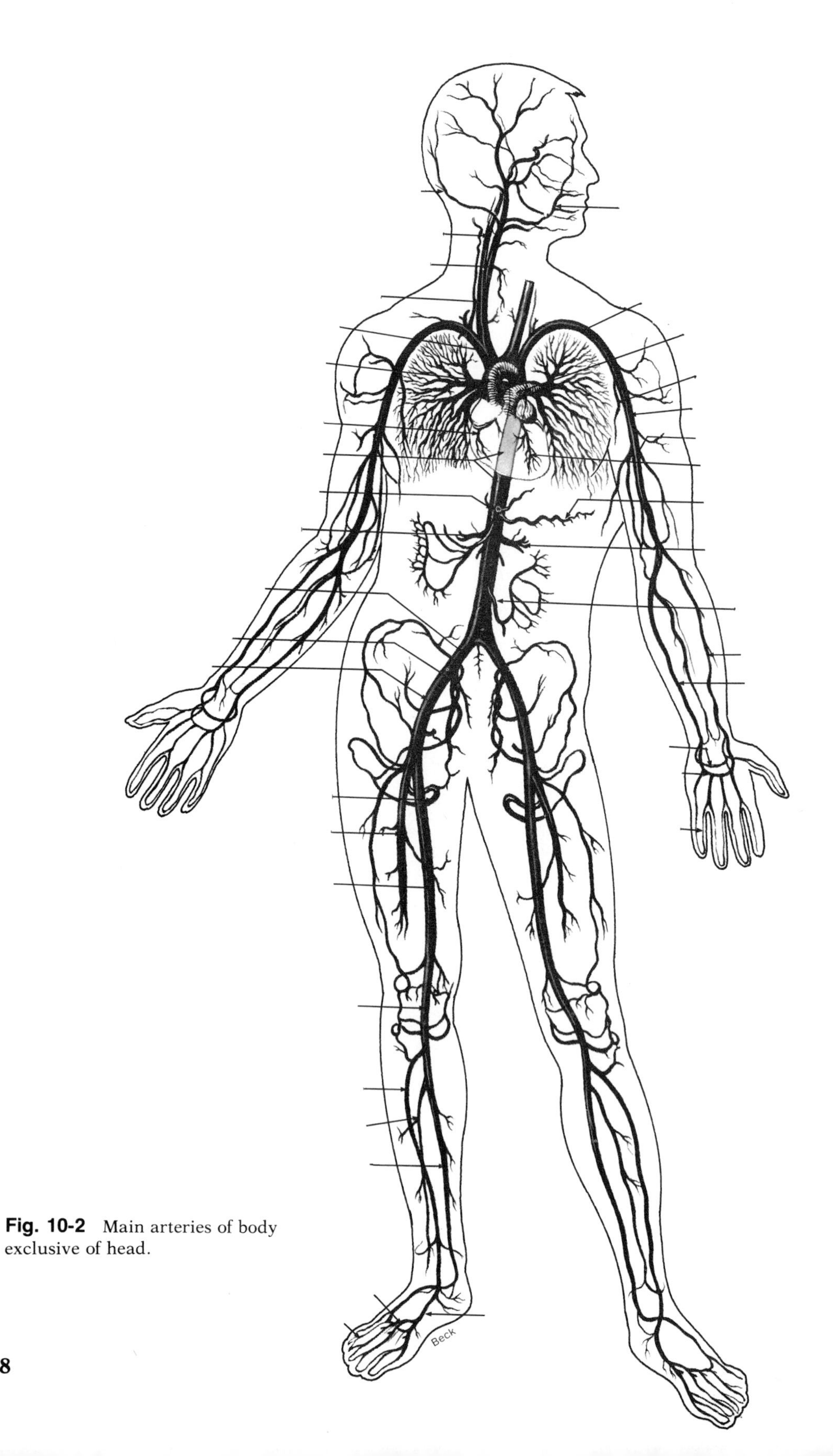

Fig. 10-2 Main arteries of body exclusive of head.

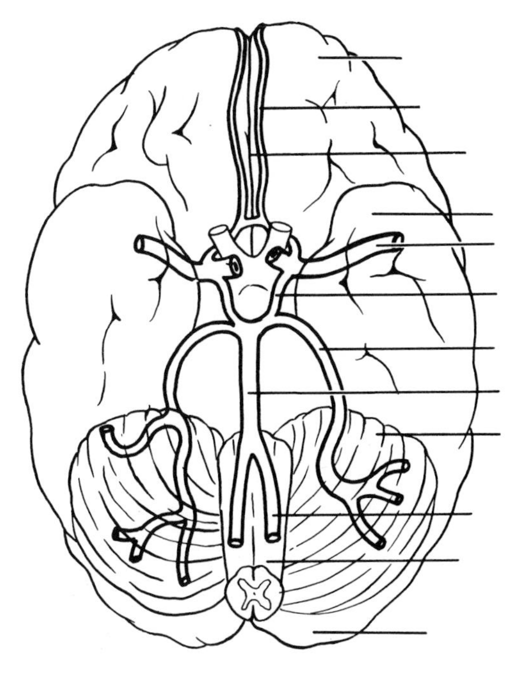

Fig. 10-3 Arterial circulation to base of brain (circle of Willis). Place each of the following terms opposite the appropriate label line:

Anterior cerebral artery
Basilar artery
Cerebellum
Posterior communicating artery
Frontal lobe of cerebrum
Longitudinal fissure
Middle cerebral artery
Occipital lobe of cerebrum
Posterior cerebral artery
Spinal cord
Temporal lobe of cerebrum
Vertebral artery

■ ■ ■

Note that arterial blood can reach the circle of Willis by two different pathways. Name the vessels that constitute each of these arterial routes from the heart to the brain.

Route 1	*Route 2*

■ Blood vessels—cont'd

Procedure B—Identification of main veins of body

Equipment

1 Charts and illustrations
2 Torso model of the human body
3 Skull
4 Textbook

Problem

What are the names and locations of the main veins of the body?

Collection of data

1 Examine textbook and chart illustrations and the torso model to identify the following veins by name and location:
Anterior tibial
Axillary
Basilic
Cephalic
Common iliac
Dorsal venous arch
External iliac
External jugular
Great saphenous
Hepatic
Inferior mesenteric
Inferior vena cava
Internal iliac (hypogastric)
Internal jugular
Left coronary
Left innominate
Medial basilic
Peroneal
Popliteal
Portal
Posterior tibial
Pulmonary
Right coronary
Right innominate
Splenic
Subclavian
Superior mesenteric
Superior sagittal sinus
Superior vena cava
Transverse sinus
2 Examine illustrations and skull to identify the veins shown on Fig. 10-6.

Conclusions

1 Place the names of the veins identified in step 1 under collection of data on their appropriate label lines on Fig. 10-4, leaving the other label lines blank.
2 Label Fig. 10-5 as directed.
3 Fill in the following blanks:
By "portal system" is meant the group of veins that drain blood from

a ______________________________

____

____

____

into the liver before it is returned to the heart. The portal vein is formed by the union of the

b ______________________________
and

c ______________________________
veins. The portal vein enters the

d ______________________________
and there branches into minute vessels (sinusoids). Arterial blood is carried into this organ by the

e ______________________________
artery. Blood is drained away from this organ by the

f ______________________________,
which empties into the

g ______________________________.

4 Label Fig. 10-6 as directed.
5 Into what vessels do the internal jugular veins drain?

____

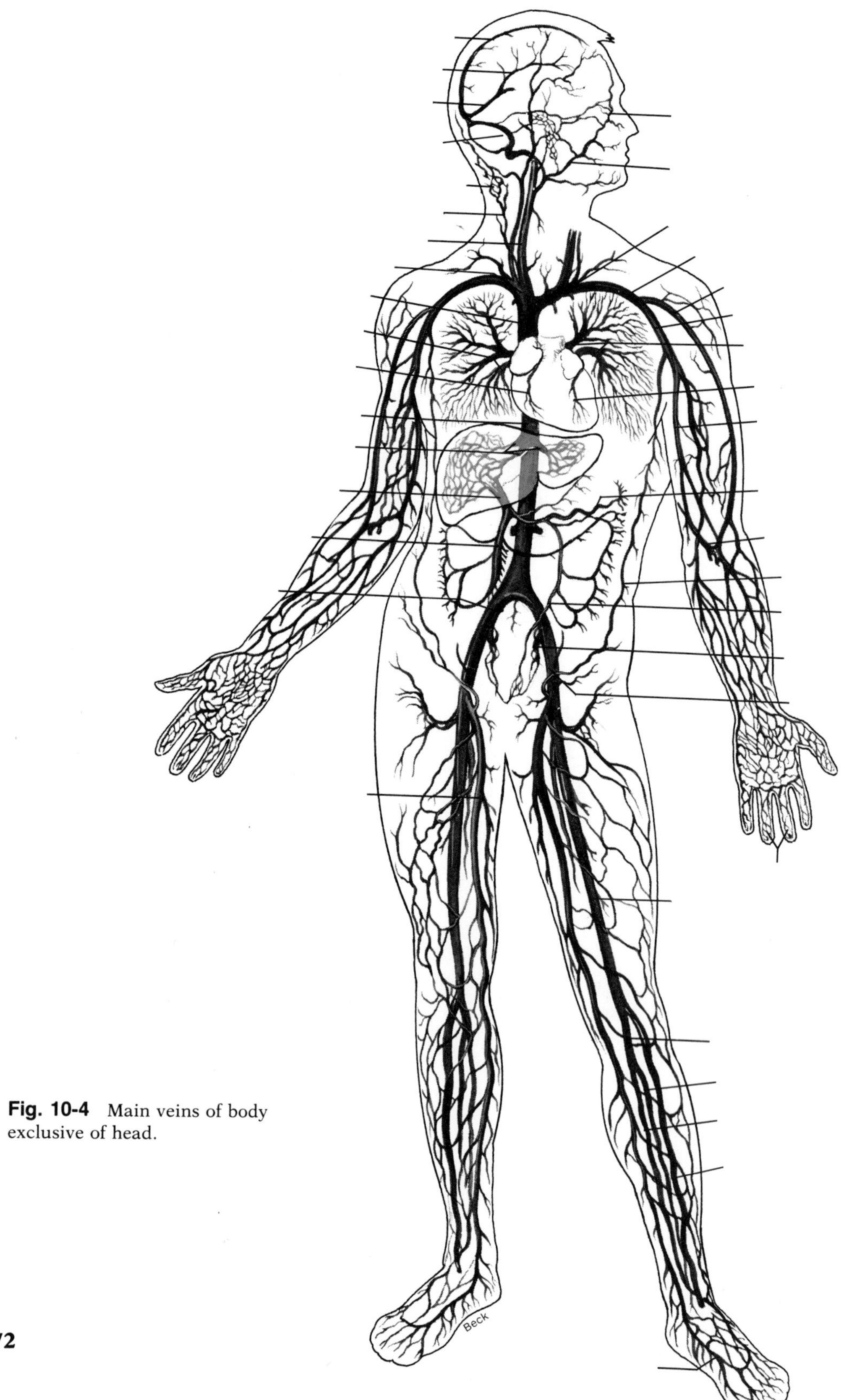

Fig. 10-4 Main veins of body exclusive of head.

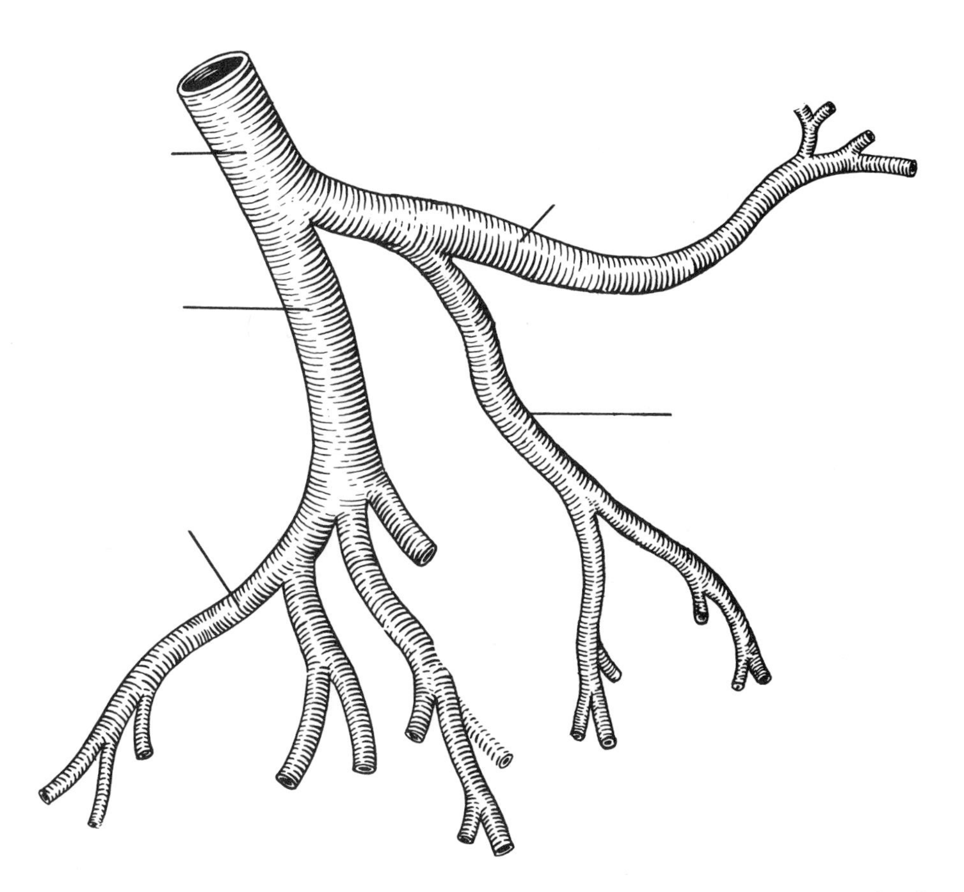

Fig. 10-5 Main tributaries of portal vein. Place each of the following terms opposite the appropriate label line:

Inferior mesenteric vein
Portal vein
Splenic vein
Superior mesenteric vein
Right colic vein

Indicate locations of the liver, spleen, and small intestine.

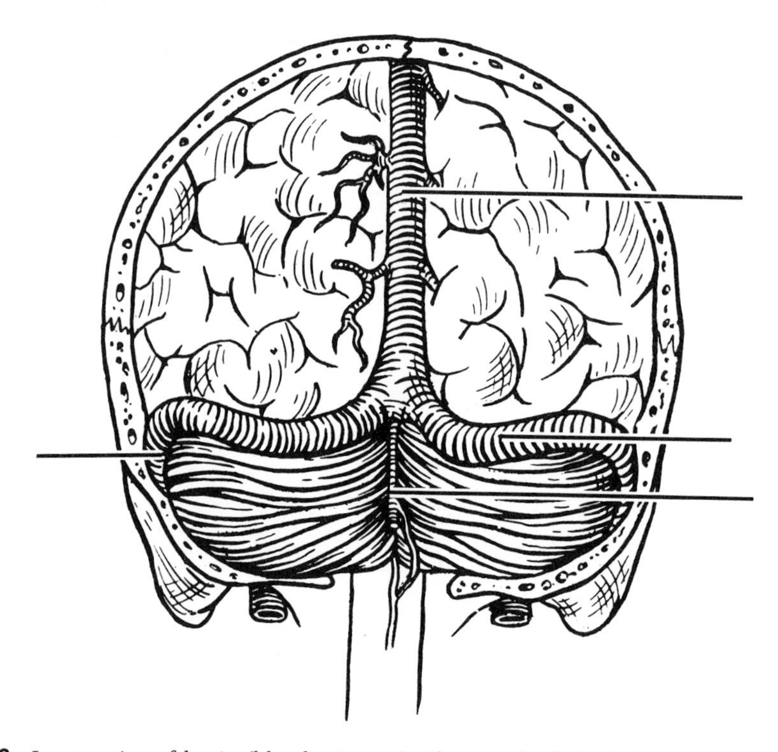

Fig. 10-6 Large veins of brain (bloody sinuses). Place each of the following terms opposite the appropriate label line:

Occipital sinus
Superior sagittal sinus
Transverse sinus

■ Blood vessels—cont'd

Procedure C—Observation of capillary blood flow

Equipment

1 Frog
2 10% urethan
3 Piece of cork (or linoleum or cardboard)
4 Microscope
5 Tincture of cantharides

Problems

1 What kind of tissue composes the capillary membrane?
2 What is the function of capillaries?
3 How does the structure of the capillary membrane relate to capillary function?

Collection of data

1 Weigh a frog. Then anesthetize it by injecting the proper amount of 10% urethan into the ventral lymph sac. Use 0.7 ml for a frog weighing 40 gm.
2 Pin the web of one foot over a small hole in a piece of cork (or linoleum or heavy cardboard) measuring about 3½ × 5 inches.
   a Identify a capillary by its small size compared with the other vessels. Observe capillary blood flow to note whether it is pulsating, steady, or irregular. Observe the manner in which red cells pass through the capillaries.
   b Change to high power and note the comparison between the diameter of red cells and that of the capillary lumen.
   c Produce inflammation by applying a small drop of tincture of cantharides to the web of the frog's foot. Watch under low power for gathering of leukocytes at the site, for migration of white and red cells out of the vessel into the tissue spaces, and for any change in the width of small vessels or in the rate of blood flow.

Conclusions

1 Capillary blood flow in the frog's foot appeared to be pulsating? steady? irregular?

____

2 The diameter of red blood cells did? did not? seem to be large enough for them to have to squeeze their way through the capillaries.

____

3 During the inflammation produced by the application of the tincture of cantharides to the frog's foot, what change, if any, did you observe in the width of the small blood vessels and in the rate of blood flow?

____

Blood pressure

Procedure A—Experiment to show whether or not a change in arterial blood pressure is followed by a change in heart rate

Equipment

1. Sphygmomanometer
2. Stethoscope
3. Watch with second hand

Problem

Is a change in arterial pressure followed by a change in heart rate, and, if so, in which direction?

Collection of data

1. Place the blood pressure cuff in position on your partner's arm. Ask her to lie down and remain quiet for 5 minutes. At the end of this time, count her pulse, measure her blood pressure, and make a note of each.
2. Ask the subject to suddenly stand up. Immediately measure and record her blood pressure.
3. About 30 seconds later, count her pulse and make a note of it.
4. About 2 minutes later, measure her blood pressure, count her pulse, and make a note of each.
5. Consult a textbook to find out what the carotid sinus is, where it is located, and what kind of receptors it contains.

Conclusions

1. What is the carotid sinus and where is it?

   ____

2. What kind of receptors are located in the carotid sinus? What kind of change stimulates them?

   ____

   ____

3. When your partner suddenly stood up from the reclining position, do you deduce that the volume of blood and therefore the blood pressure in her carotid sinus increased or decreased? Why?

   ____

   ____

4. Did the data you collected in this experiment indicate that a change in arterial pressure is followed by a change in heart rate?

   ____

5. Did you find that the heart rate increased or decreased following the sudden decrease in arterial blood pressure because of the change from the reclining to upright position?

   ____

6. Describe the homeostatic mechanism thought responsible for the changes observed in the foregoing experiment by encircling the appropriate terms in parentheses and filling in the blank line before the word "receptors" in Fig. 10-7.
7. Based on the data you collected in the experiment, how would you expect a patient's heart rate to change soon after he had had a hemorrhage?

   ____

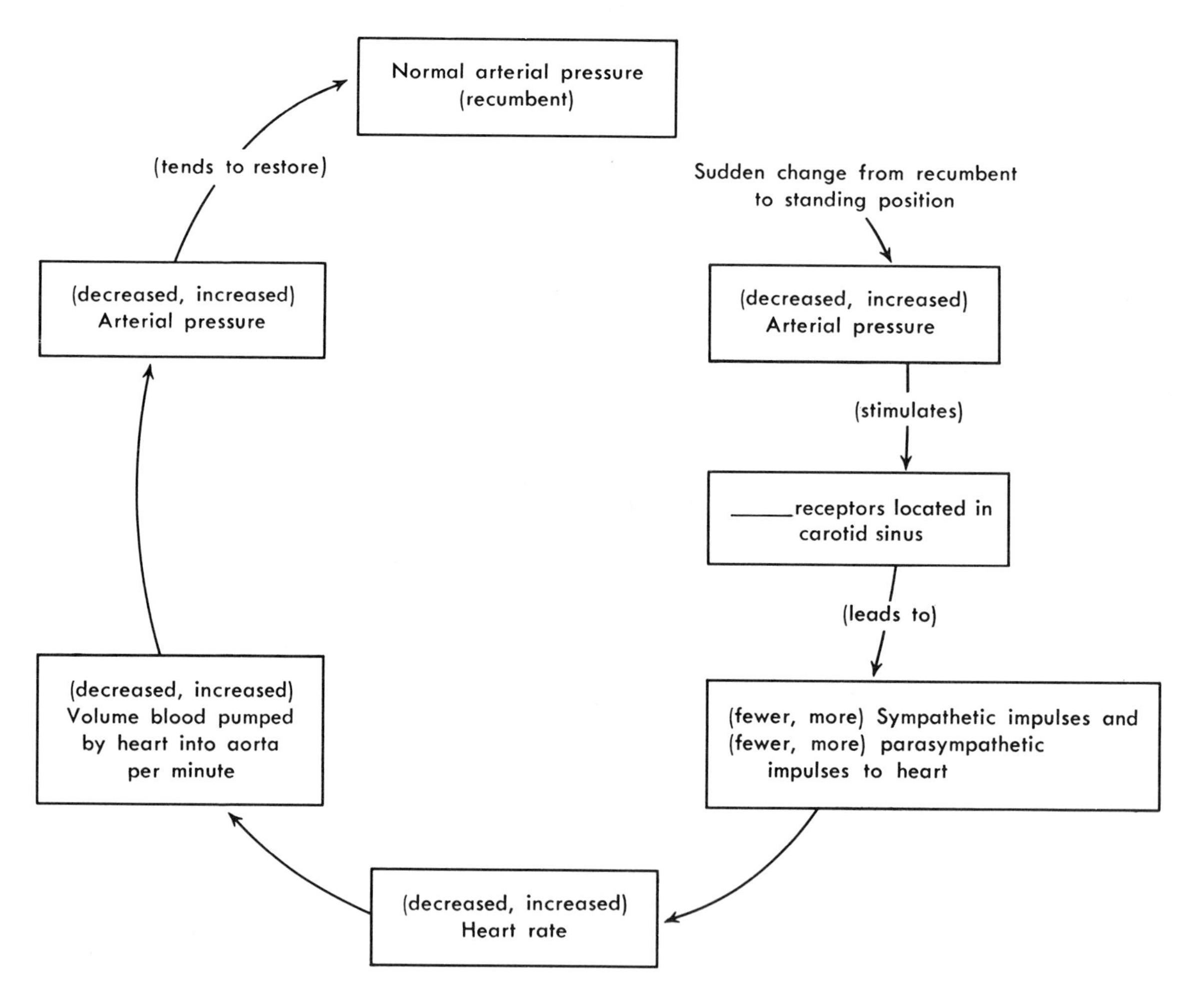
Normal arterial pressure
(recumbent)
(tends to restore)
Sudden change from recumbent
to standing position
(decreased, increased)
Arterial pressure
(decreased, increased)
Arterial pressure
(stimulates)
_____receptors located in
carotid sinus
(leads to)
(decreased, increased)
Volume blood pumped
by heart into aorta
per minute
(fewer, more) Sympathetic impulses and
(fewer, more) parasympathetic
impulses to heart
(decreased, increased)
Heart rate

Fig. 10-7

■ Blood pressure—cont'd

Procedure B—Estimation of normal or abnormal venous blood pressure

Equipment

Mirror large enough to reflect upper half of body

Problem

How can one estimate whether venous pressure is normal or not?

Collection of data

1 Sit in a chair with your hands resting on your lap. Observe whether the veins on the backs of your hands appear distended or flat.
2 Raise one hand up to the level of your suprasternal notch. Observe whether hand veins appear distended or flat.
3 Sit in front of a mirror with your head at a 30 to 45 degree angle. Observe whether your neck veins appear distended or flat.
4 If venous pressure is normal, the veins on the backs of the hands appear distended when an individual is sitting with hands resting in lap, but they collapse when hand is raised to level of suprasternal notch.
5 If venous pressure is normal, neck veins appear flat when individual is sitting with head at 30 to 45 degree angle.
6 If venous pressure is higher than normal, hand veins do not collapse when hand is raised to level of suprasternal notch, and neck veins appear distended when individual is in sitting position with head at 30 to 45 degree angle.

Conclusions

1 From your observations in steps 1 and 2 under collection of data, do you estimate that you have a normal or higher than normal venous pressure?

____

2 From your observation in step 3, do you estimate that your venous pressure is normal or higher than normal?

____

Blood pressure—cont'd

Procedure C—Effect of the Valsalva maneuver on central venous pressure and on the volume of blood returning to the heart

Equipment

Mirror

Problems

1 What effect does the Valsalva maneuver have on central venous pressure (i.e., pressure in the venae cavae)?
2 What effect does the Valsalva maneuver have on the volume of blood returning to the heart?

Collection of data

1 Observe your neck veins (jugular) in a mirror while you perform the Valsalva maneuver. To do this, hold your breath, fix your thorax (keep it stationary), and at the same time forcibly contract your abdominal muscles. This action, the Valsalva maneuver, is commonly called "straining," as, for example, in defecation.
2 Maintain the Valsalva maneuver a few seconds, then, while watching your neck veins in a mirror, suddenly exhale.
3 Consult a textbook to confirm the name of the veins into which the jugular veins drain.

Conclusions

1 What change did you observe in your neck veins while you were performing the Valsalva maneuver?

____

2 Did this change in the appearance of your neck veins indicate an increase or a decrease in the volume and pressure of the blood in them?

____

3 Considering that the abdominal muscles contract forcibly but that expiration is prevented by holding the breath during the Valsalva maneuver, do you deduce that this maneuver increases or decreases intrathoracic pressure and therefore central venous pressure?

____

It is this change in intrathoracic and central venous pressure that causes the changes described in conclusions 1 and 2.

4 Do you conclude that the change in intrathoracic pressure during the Valsalva maneuver increases or decreases the volume of blood draining out of the jugular veins (and other peripheral veins also) into the central veins?

____

5 Judging from the rapid change in appearance you observed in your neck veins when you terminated the Valsalva maneuver by suddenly exhaling, do you conclude that the drainage of blood from the neck veins into the central veins and on into the heart markedly increases or decreases upon termination of the Valsalva maneuver? Does this increase or decrease the work load put on the heart? For this reason, patients with a weak or diseased heart should? should not? perform the Valsalva maneuver.

____

____

____

Exercise and circulation

Procedure—Film

"Running for life," color, 16 mm, 28 min; Indiana University, Audio-Visual Center, Bloomington, Ind. 47401 (purchase, $240; rental, $10). Based on 2-year experiment sponsored by U.S. Office of Public Health to determine effects of exercise on middle-aged persons and whether exercising can lessen the chances of heart disease.

Self-test

1 Blood clotting
Describe briefly the main steps in blood clotting.

____

____

____

____

____

____

2 Trace the flow of blood through the heart.

____

____

____

____

____

____

____

____

____

____

____

____

____

____

3 Systemic blood flow
Name the vessels in the order in which blood flows through them to reach the capillaries of the small intestine from the left ventricle of the heart and to return to the right atrium of the heart.

____

____

____

____

____

____

____

____

____

____

____

____

4 An important principle to remember is that a sudden decrease? increase? in arterial blood pressure initiates reflex acceleration of the heartbeat.

____

5 One cardinal sign of hemorrhage is a marked drop in the arterial blood pressure. Another cardinal sign, therefore, is a slow? rapid? pulse.

____

6 Restating the principle formulated in question 4: as a rule, when the blood pressure decreases, the pulse decreases too? increases?

____

11 The digestive system and metabolism

■ Digestive organs and deglutition

Procedure A—Demonstration of digestive organs of anesthetized animal

Equipment

1 Rat (nothing to eat for 24 hours; then about 2 hours before laboratory period, feed cream and a little broth)
2 Ether
3 Cotton
4 Gauze sponges
5 Dissecting instruments and pans
6 Physiological saline solution

Problems

1 What are gross structural features of the digestive system?
2 What is the nature of stomach and intestinal movements?
3 Into what structures are fats absorbed?

Collection of data

1 Study textbook or chart illustrations of the digestive tract. Then follow directions given under conclusion.
2 Anesthetize the animal by placing it under a bell jar with ether-saturated cotton sponges. When the excitement stage has passed, prod the animal with the side of the jar or with a blunt object. If it does not respond, it is sufficiently anesthetized. Keep ether and cotton ready for further anesthetizing if the animal should start moving during dissection.
3 Make a median incision from the sternum to the pubis, cutting through the skin and fascia. Slit the peritoneum carefully to avoid injuring the underlying structures.
4 Examine the omentum, noting what kind of membrane composes it.
5 Examine the stomach, noting its size, shape, curvatures, and movements.
6 Lift up the omentum. Place gauze sponges soaked in hot physiological saline solution over a section of intestine. Observe the intestines for movements.
7 Compare small and large intestines as to length and diameter. Note movements.
8 Examine the mesentery carefully. Look for lymphatics that appear milky white from fat absorbed into them.
9 Examine the liver, bile ducts, pancreas, and spleen. Note whether or not the rat has a gallbladder.
10 Cut open the stomach along the lesser curvature and continue the incision down into the small intestine. Wash out the interior of these organs. Observe the rugae in the stomach and the circular folds (through magnifying glass) in the intestine.
11 Kill the animal by bleeding.

Suggested demonstrations

1. If possible, arrange for the students to watch a fluoroscopic examination of the gastrointestinal tract.
2. Demonstrate bile ducts on sheep liver (with gallbladder attached) obtained from a slaughterhouse.

Digestive organs and deglutition—cont'd

Procedure B—Demonstration of deglutition

Equipment

1 Stethoscope
2 Glass of water

Problems

1 What part does the tongue play in swallowing?
2 What part does the larynx play in swallowing?

Collection of data

1 Drink a glass of water slowly, observing as you do so whether or not the tongue moves and, if so, in which direction.
2 Place your fingers gently on your larynx and swallow some more water. Note its movements.
3 Take a firm hold of your larynx and try to prevent its moving while you try to swallow.

Conclusions

1 Did your tongue move as you swallowed? If so, in which direction did it seem to move?

____

____

2 Were you able to prevent the larynx from moving as you swallowed?

____

3 What prevents food from entering your larynx and trachea when you swallow?

____

____

____

■ Digestive process
Procedure—Films

"Human digestion," color, 16 mm, 10 min; McGraw-Hill Films, Dept. WP, 330 West 42nd St., New York, N. Y. 10036 (purchase, $75; rental, $10).

or

"The human body; the chemistry of digestion," color, 16 min; Modern Film Rentals, 2323 New Hyde Park Road, New Hyde Park, N. Y. 11040 (rental, $11).

or

"The human body; digestive system," color, 13½ min; Modern Film Rentals, 2323 New Hyde Park Road, New Hyde Park, N. Y. 11040 (rental, $8). Includes live action scenes of major digestive organs and detailed account of functions of various organs of this system.

■ Metabolism
Procedure—Films

"How the body uses energy," color, 16 mm, 15 min; McGraw-Hill Films, Dept. WP, 330 West 42nd St., New York, N. Y. 10036 (purchase, $190; rental, $12.50).

and/or

"Energy cycles in the cell," color, 16 min; McGraw-Hill Films, Dept. WP, 330 West 42nd St., New York, N. Y. 10036 (purchase, $215; rental, $12.50).

12 The urinary system

Kidney structure

Procedure A—Gross structure: dissection of sheep kidneys

Equipment

1. Sheep kidneys
2. Dissecting instruments
3. Dissecting pans
4. Toothpicks
5. Models of torso and kidney circulation

Problem

What gross structural features characterize the kidneys and other urinary organs?

Collection of data

1. Study Figs. 12-1 to 12-3 and the torso model.
2. Before cutting into a sheep kidney, observe the large amount of fat encasing it.
3. a Remove the fat from around the kidney, using care not to tear away the vessels that enter the kidney. Look for a small gland embedded in the fat around the superior end of the kidney.
   b Observe the shape, color, and texture of the kidney. Note the slit on the concave surface through which the vessels enter the kidney.
4. Examine the shiny, transparent covering membrane adherent to the kidney. Remove a piece of it with a forceps.
5. Find the ureter, renal artery, and renal vein. The latter is the thinnest-walled vessel entering the kidney. Therefore, it does not stand open but is collapsed. The ureter has the thickest wall of the three vessels and is the largest. Insert toothpicks into each of these structures.
6. Holding the kidney in your left hand, with convex side outward, use a sharp scalpel to make a coronal section, but do not separate the two halves completely. Lay the open organ on the tray. Observe the points of entrance of the vessels into which you inserted the toothpicks.
7. Identify the following structures:
   a Renal capsule
   b Renal cortex
   c Renal medulla
   d Renal pyramids
   e Renal papilla
   f Renal pelvis
   g Renal calyces
8. Identify branches of the renal artery and vein on the model. Cut away the renal pelvis. Scrape off the underlying adipose tissue to expose the branches of the renal artery. Trace several of these.

Conclusions

1. The strong, tough, transparent membrane adherent to the outer surface of the kidney is called the renal ....

   ____

2. Mucous? Serous? Synovial? membrane lines the renal pelvis and calyces.

   ____

3 The ureters are mucosa-lined tubes that connect the . . . . with the . . . .

---

4 The mucosa-lined tube through which urine moves from the bladder to the exterior is named the . . . .

---

5 Fig. 12-2 shows that in the female, the opening into the urethra lies anterior and slightly superior to the opening into the . . . .

---

6 As you can see in Fig. 12-3, the outer portion of the kidney, lying immediately under the renal capsule, is known as the . . . .

---

7 The medulla is the inner? outer? part of the kidney and it consists of pyramids.

---

8 As you can see in Fig. 12-3, the pelvis of the kidney is the expanded upper end of the . . . .

---

9 Note that both the renal pelvis and its divisions, each of which is called a . . . . . . , lie inside the kidney.

---

10 Careless technique in catheterizing may introduce microbes into the bladder. It could cause an infection of the bladder that would spread to ureters, pelvis, and calyces since a continuous sheet of . . . . . membrane lines all of these structures.

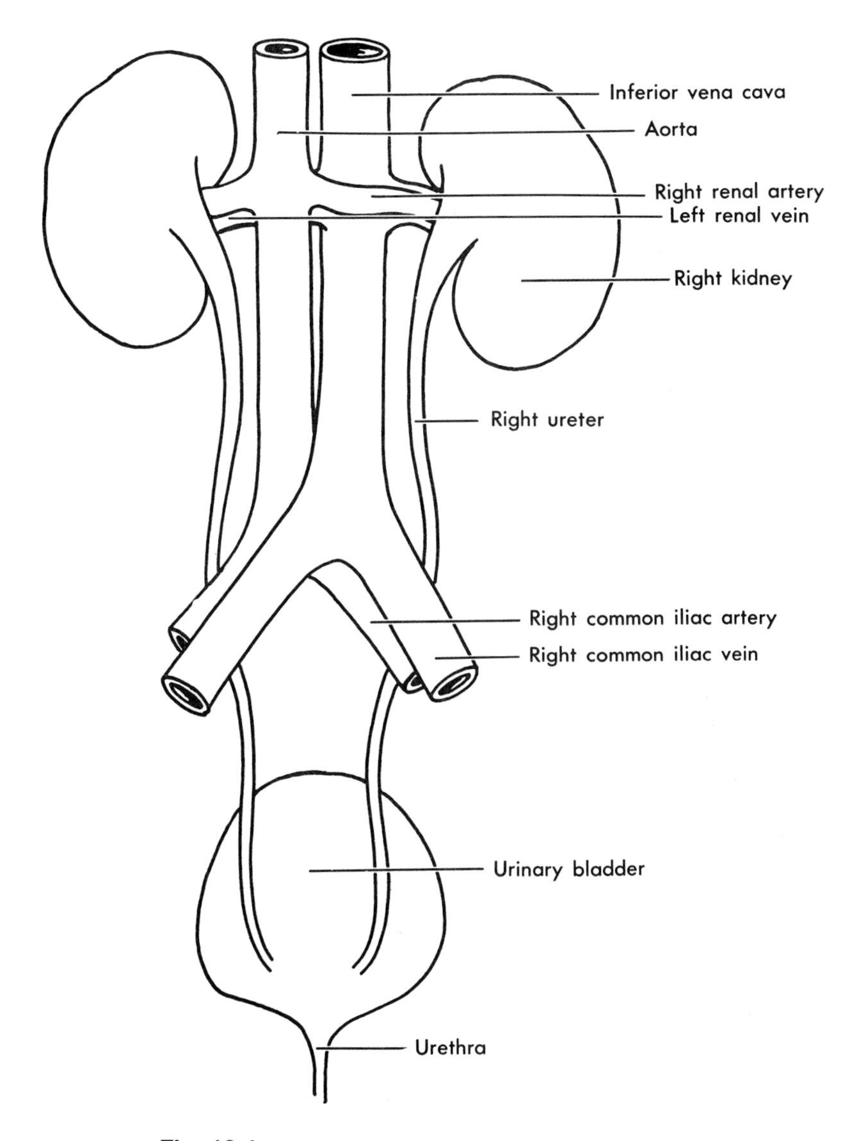

Fig. 12-1 Urinary organs viewed from behind.

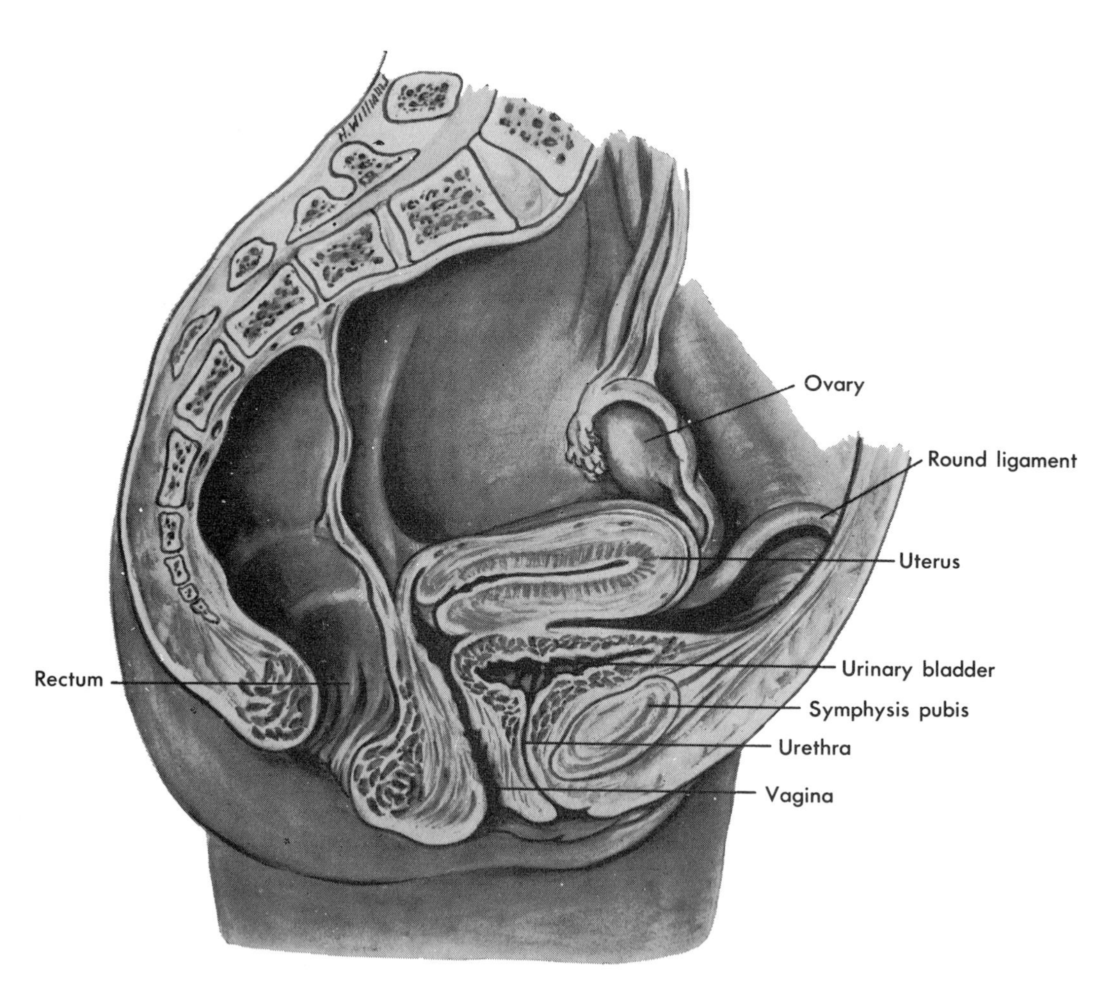

Fig. 12-2 Sagittal section through midline of female pelvis.

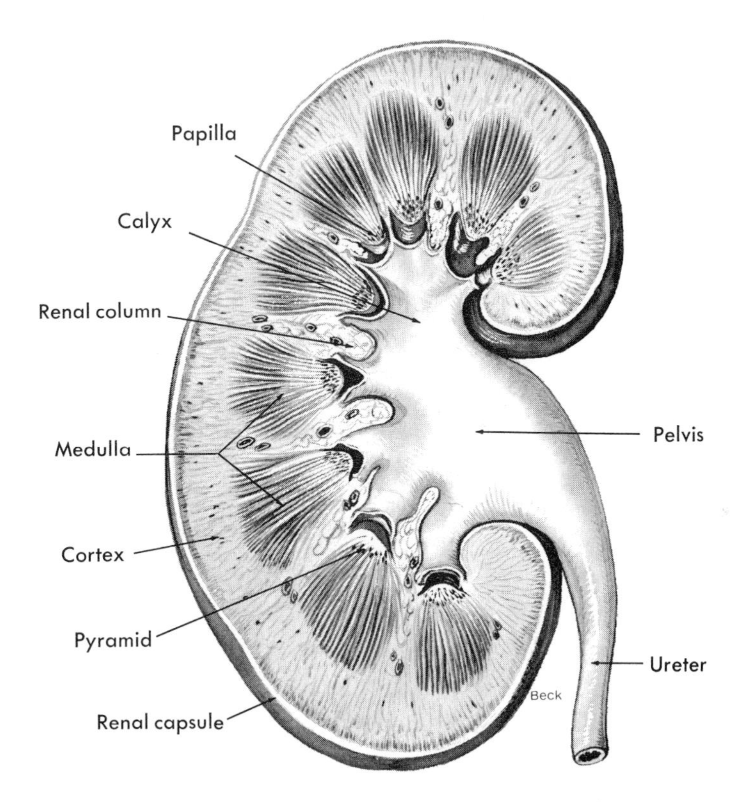

Fig. 12-3 Coronal section through right kidney.

■ Kidney structure—cont'd
Procedure B—Microscopic structure of kidneys

Equipment

1 Microscopic or projection slides of kidneys
2 Illustrations
3 Textbooks on anatomy, physiology, and histology

Problems

1 What is a nephron?
2 What are its structural characteristics?
3 What three functions do different parts of a nephron perform?

Collection of data

Study Fig. 12-4. Then consult data it contains to answer items under conclusions.

Conclusions

1 Invaginated in Bowman's capsule shown in Fig. 12-4 is a network of capillaries called the ....

___

2 A nephron is one of the million or more functional units of each kidney. Nephrons actually form the urine excreted by the kidney. One nephron consists of a glomerulus (i.e., a ....), a .... around the glomerulus, and a tubule emerging from Bowman's capsule.

___

___

3 Since a million or more nephrons constitute each kidney, do you think you could see a nephron without the help of a microscope?

___

4 Note the shape of a renal tubule (tubule attached to Bowman's capsule). The straight, most distal portion of it is called a ....

___

5 The proximal convoluted tubule is the coiled part of a renal tubule located farther from? nearer to? Bowman's capsule than the .... convoluted tubule.

___

___

6 Between the proximal and distal convoluted tubules lie the descending limb and ascending limb of ....

___

7 As you can deduce from Fig. 12-4, blood enters the glomerulus by way of an ..... arteriole and leaves it by way of an ..... arteriole.

___

___

8 The first step in urine formation is this: fluid and true solutes (but normally not colloidal solutes) filter out of blood flowing slowly through the .... into Bowman's capsule.

___

9 As the filtrate moves out of Bowman's capsule down the .... tubule and through the limbs of the loop of Henle, the second step in urine formation takes place—viz., the reabsorption of water, ions, glucose, and other nutrients. Reabsorption means that these substances move out of blood? out of tubular filtrate? into ....

___

___

___

10 The final step in the nephron's formation of urine goes on in the distal convoluted and collecting tubules. It consists of secretion, mainly of potassium and/or hydrogen ions in exchange for the reabsorption of sodium ions. Summarizing briefly, urine formation consists of: .....

from glomerulus into Bowman's capsule; reabsorption and? but not? secretion from proximal tubules and loops of Henle; and secretion and? but not? reabsorption from distal and collecting tubules.

____

____

____

■ Kidney function

Procedure—Film

"Functional anatomy of the human kidney," color, 16 mm, 32 min (with the aid of microphotography); Smith Kline & French Laboratories, 1500 Spring Garden St., Philadelphia, Pa. 19101 (sponsor—Smith Kline & French Laboratories).

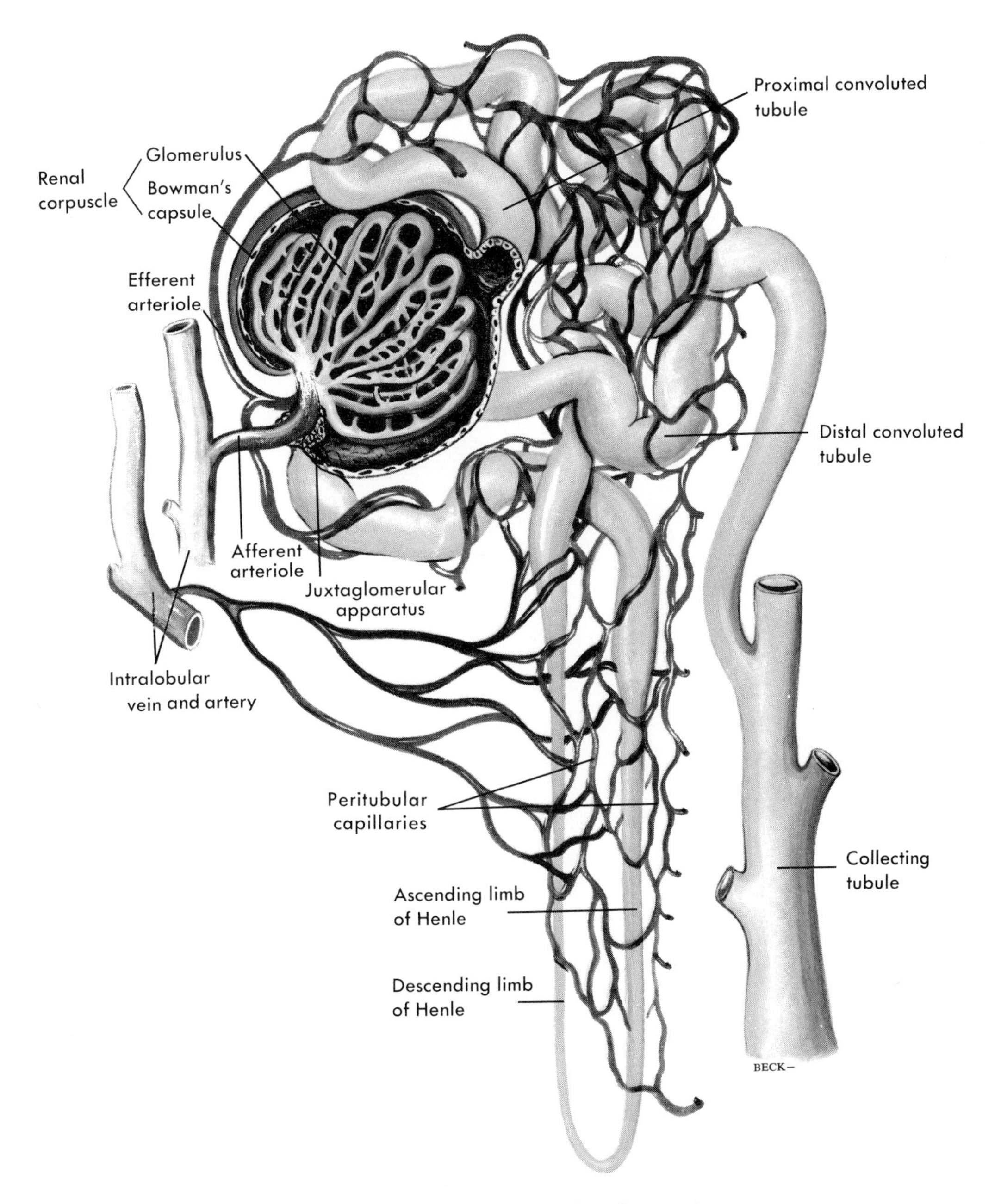

Fig. 12-4 Blood vessels of nephron unit.

UNIT FIVE

REPRODUCTION

13 Reproduction of cells

DNA and mitosis
Procedure—Films

"DNA," color, 11 min; McGraw-Hill Films, Dept. WP, 330 West 42nd Street, New York, N. Y. 10036 (purchase, $160; rental, $12.50). Introduces concept of DNA through animated diagrams, three-dimensional models, and photomicrographs; illustrates DNA molecule in process of copying itself during mitosis.

"Reproduction in cells (the 5 phases of mitosis)," color, 16 min; McGraw-Hill Films, Dept. WP, 330 West 42nd St., New York, N. Y. 10036 (purchase, $215; rental, $12.50).

"Functions of DNA and RNA," color, 13½ min; Modern Film Rentals, 2323 New Hyde Park Road, New Hyde Park, N. Y. 11040 (rental, $11).

Meiosis
Procedure—Film

"Chromosomes and genes (meiosis)," color, 16 min; Modern Film Rentals, 2323 New Hyde Park Road, New Hyde Park, N. Y. 11040 (rental, $11).

14 Reproduction

Male reproductive system

Procedure—Gross structure of male reproductive organs

Equipment

1 Embalmed male cat
2 Lamb scrotum from slaughterhouse
3 Charts and illustrations
4 Model of male pelvis
5 Dissecting instruments
6 Dissecting pans

Problem

What are the major gross structural features of the male reproductive system?

Collection of data

1 Study Fig. 14-1 and examine a model of the male pelvis, noting the locations and appearances of the following:
   a Bulbourethral glands
   b Vas deferens
   c Ejaculatory ducts
   d Epididymis
   e Penis
   f Prostate
   g Scrotum
   h Seminal vesicles
   i Spermatic cords
   j Testes
2 Locate as many of these structures as you can on an embalmed male cat.
3 Dissect a lamb scrotum, identifying the seminiferous tubules, epididymis, and beginning part of the vas deferens.

Conclusions

1 As you learned in Chapter 12, the mucosa-lined tube by which urine leaves the body is the ....

____

2 The male reproductive fluid (semen) moves out of the ejaculatory duct into the ...., the tube by which it leaves the body.

____

3 The urethra serves as the duct to the exterior for the urinary tract and? but not? the reproductive tract in the male.

____

4 In the female, as Fig. 12-2 (p. 194) shows, the urethra serves as the duct to the exterior for the urinary tract and? but not? the reproductive tract.

____

5 Since the prostate gland is a doughnut-shaped gland, you can deduce from Fig. 14-1 that the .... passes through the center of this gland.

____

6 If a man's .... gland becomes sufficiently swollen, it decreases, or may even

obstruct entirely, urine flow out of the bladder to the exterior.

---

7 Ducts from the testis merge to form a single, tightly coiled tube, called the . . . ., which lies across the top and sides of the testis.

---

8 A tube from the right and from the left epididymis in the scrotum extends into the pelvic cavity; this extension, as Fig. 14-1 shows, is called the . . . ., and the surgical procedure that severs both of these tubes is called a . . . . . . -ectomy.

---

9 After entering the pelvis, each vas deferens curves over the top of the bladder and on its posterior surface joins a duct, as you can see in Fig. 14-1, from the . . . . to form the . . . . duct, which in turn joins the . . . .

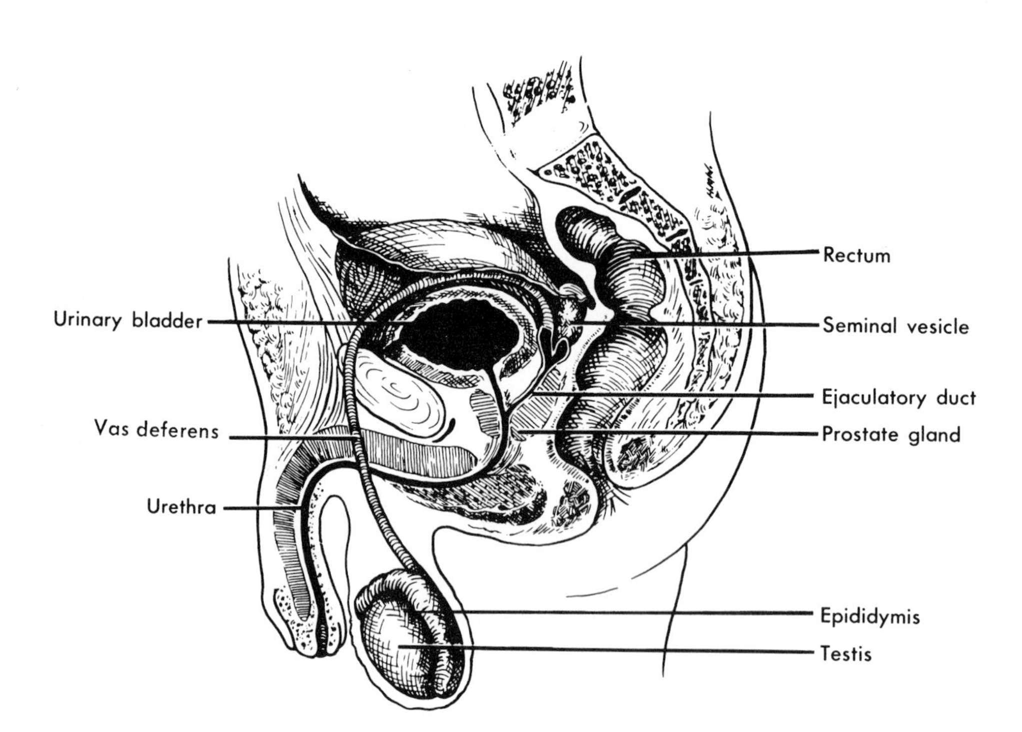

Fig. 14-1 Sagittal section through male pelvis.

Female reproductive system

Procedure—Gross structure of female reproductive organs

Equipment

1 Embalmed female cat (preferably pregnant)
2 Model of female pelvis

Problem

What major gross structural features characterize the female reproductive system?

Collection of data

1 Study Fig. 12-2 (p. 194) and examine the model of the female pelvis, charts, and illustrations, noting the locations and the appearance of the following:
   a Ovaries
   b Uterus and its ligaments
   c Uterine tubes
   d Vagina
2 Locate and examine these structures on the embalmed female cat.

Conclusions

1 The two male sex glands, the ...., lie outside the pelvic cavity in the scrotum.

____

2 As Fig. 12-2 (p. 194) shows, the two female sex glands, the ovaries, lie inside? outside? the .... cavity.

____

____

3 Just as the testes form the male sex cells (sperm), the .... form the female sex cells (ova).

____

4 The largest female reproductive organ shown in Fig. 12-2 (p. 194) is the ......, the organ in which an impregnated ovum develops into a baby.

____

5 The terminal (distal) portion of the male reproductive tract is the ...., whereas the distal portion of the female tract is the ....

____

____

Sex cells and fertilization

Procedure—Films

"The egg and sperm," color, 16 mm, 16 min; McGraw-Hill Films, Dept. WP, 330 West 42nd St., New York, N. Y. 10036 (purchase, $215; rental, $12.50).

"The fertilization process," color, 16 min; McGraw-Hill Films, Dept. WP, 330 West 42nd St., New York, N. Y. 10036 (purchase, $215; rental, $12.50).

Reproductive hormones

Procedure—Films

"Reproductive hormones," color, 16 min; McGraw-Hill Films, Dept. WP, 330 West 42nd St., New York, N. Y. 10036 (purchase, $215; rental, $12.50).

"A changing view of the change of life," color, 28 min; sponsored by Wilson Research Foundation, New York, N. Y.; prepared under supervision of R. A. Wilson, M.D., and E. R. Marino, M.D.; produced by Dynamic Films, Inc., 1965 (free loan from Association Films, 600 Madison Ave., New York, N. Y. 10022). New medical attitudes toward female menopause, presented as a deficiency disease brought on by hormonal imbalance.

"The mechanisms of action of oral contraceptives," color, 21 min; loan information from Professional Services, Syntex Laboratories, Inc., 3401 Hillview Ave., Palo Alto, Calif. 94304.

"Is the pill dangerous?" color, 28 min; Lawren Productions, P.O. Box 1542, Burlingame, Calif. 94010 (purchase, $225; rental, $20).

Embryology

Procedure—Embryological development

Equipment

1 Three dozen fertile eggs
2 Incubator

Problem

What are some of the major characteristics of embryological development?

Collection of data

1 Twenty-one days before the laboratory period during which this demonstration is to take place, put one fertile egg in an incubator. Put an additional egg in each day thereafter for the next 21 days. Turn each egg each day. Keep the temperature at 38° C. Keep a water-filled pan in the incubator.
2 Open all of the eggs in the laboratory 21 days after the first egg was placed in the laboratory.
3 Also display models or illustrations of human embryos of different ages. Students compare chick specimens with human embryos.

Self-test

Test yourself by answering questions on pp. 210-211 and by completing items listed here.

1 One of the main functions of the ovaries is the secretion of two hormones named .... and ....

2 The other main function performed by the ovaries is ....

3 The only ovarian structure that secretes progesterone is the one called the ....

4 Graafian follicles secrete most of the .... produced by the ovaries.

5 The corpus luteum secretes progesterone and also? but not? estrogens.

6 Graafian follicles secrete estrogens and also? but not? progesterone.

7 Each ovary actually functions as two different endocrine glands—viz., the .... and ....

8 Besides secreting estrogens, graafian follicles also serve another function—viz., they produce ....

9 Hormones known collectively as gonadotropins and secreted by the .... gland control the ovaries' secretion of hormones and its production of ova.

10 Two gonadotropins secreted by the anterior pituitary gland are .... and ....

11 FSH is the abbreviation for ....

12 LH is the abbreviation for ....

13 A high blood concentration of .... initiates the process by which a follicle and its enclosed ovum "ripen" (develop to maturity).

14 Blood concentration of FSH must reach a certain high level in order to initiate ....

15 If during any month the blood concentration of FSH stays at a high? low? level, no follicle or ovum begins maturing that month.

16 Whereas a high blood concentration of the anterior pituitary hormone, FSH, stimulates ovarian secretion of ...., a high blood concentration of estrogens inhibits anterior pituitary secretion of FSH.

17 If for all but a few days of any month the blood concentration of estrogens remains high, the anterior pituitary gland will secrete an excess of? little or no? FSH that month.

18 If in any month the anterior pituitary gland produces little or no ...., that month no mature ovum will develop and be released. In other words, that month ovulation will not occur.

19 One ingredient of contraceptive pills is a preparation of the hormone ...., which inhibits FSH secretion by the anterior pituitary gland.

20 In other words, contraceptive pills act directly? indirectly? on the ovaries to inhibit ovulation. One way in which contraceptive pills may act is to build up a high blood estrogen concentration that acts on the .... gland to inhibit its production of .... With a low blood concentration of this hormone, the ovaries are not stimulated to develop an ovum and they do not ovulate.

---

21 A negative? positive? feedback mechanism is said to operate between the anterior pituitary gland and the ovaries because the anterior pituitary hormone, FSH, has an inhibiting? a stimulating? effect on the ovaries, whereas the ovarian hormone, estrogen, has an inhibiting? a stimulating? effect on the anterior pituitary gland.

---

22 The abbreviation ICSH stands for the male hormone ....

---

23 ICSH, as its name suggests, stimulates interstitial cells of the testes to secrete their hormone called ....

---

24 ICSH, in the female, is called ....

STRESS

■ Selye's concept of stress
Procedure—Film

"Stress and the adaptation syndrome," color, 35 min; Pfizer Medical Film Library, 267 West 25th St., New York, N. Y. 10001 (free loan). The classic film made in 1956 in cooperation with Dr. Selye by N. P. Schenker, M.D.